A Long Look at Nature

A Long Look at Nature

The North Carolina State Museum of Natural Sciences

MARGARET MARTIN

Original Photography by Rosamond Purcell

PUBLISHED FOR THE NORTH CAROLINA

STATE MUSEUM OF NATURAL SCIENCES BY THE

UNIVERSITY OF NORTH CAROLINA PRESS

CHAPEL HILL AND LONDON

For Mary,
whose determination and strength
inspired me to get on with this
book — and other, more important,
things in life —

Margaret

Designed by April Leidig-Higgins
Set in Quadraat by Eric M. Brooks
Manufactured in Hong Kong

The paper in this book meets
the guidelines for permanence
and durability of the Committee
on Production Guidelines for
Book Longevity of the Council
on Library Resources.

**Grateful acknowledgment is
given to Friends of the Museum
for their generous support of
this project.**

**Photography was supported by
a grant from the North Carolina
Arts Council, a state agency.**

Complete cataloging data
is available from the Library
of Congress.
ISBN 0-8078-4985-5
(pbk.: alk. paper)

05 04 03 02 01   5 4 3 2 1

Page ii: Northern quahogs,
Mercenaria mercenaria; pages iv–v:
Mineral cores from nineteenth-
century North Carolina mine;
pages vi–vii: Northern quahogs,
Mercenaria mercenaria; page xii:
Fossilized sturgeon dermal armor.
All courtesy of Rosamond Purcell.

In honor of the dedicated naturalists whose work in the field and in the State Museum's collections has enhanced our appreciation of the natural world.

CONTENTS

Fresh off the boat from England in 1700, young John Lawson immersed himself in the wild country of North Carolina. The landscape, he wrote in *A New Voyage to Carolina*, is "adorn'd by Nature with a pleasant Verdure, and beautiful Flowers, frequent in no other Places . . . large and spacious Rivers, pleasant Savanna's, and fine Meadows . . . proper Habitations for the Sweet-singing Birds." Soon Lawson was sending specimens of the verdure—the rare flowers, the minerals, and the animals—of this exotic land to eager collectors in Europe. He planned "to make a strict collection" of North Carolina's beasts, birds, fishes, insects, fossils, minerals, plants, and soils. For the next two centuries, museums in Europe and America would seek out North Carolina's natural treasures as prize specimens in their collections.

By the nineteenth century, natural history collections had moved from the private domain of wealthy men and learned societies to the public sector of the natural history museum. Museums bore the cultural role of public interpreter of nature, a role now shared with television, magazines, popular films, and the Internet.

This book looks at North Carolina's natural landscape through the experience of the North Carolina State Museum of Natural Sciences, whose collections and interpretive exhibits reflect the relationship between society and nature. The emphasis of the museum's collections changed as people changed the way they valued the natural resources of their state. When the museum was founded in 1879, its collections of flora and fauna shared space

Summer tanagers, *Piranga rubra*. Courtesy of Rosamond Purcell.

photography came from Teddi Brown, Heather Kowalski, Nancy Childs, April Short, Amme Maguire, Lisa Yow, Tamara Trentlage, Roy Campbell, Sheree Worrell, and Brenda Wynn.

Many other people helped with the preparation of individual chapters, and they are acknowledged at the end of the book under "Sources."

with natural treasures that were ripe for trade—timber, stone, fisheries, and agricultural products. The State Museum became both a promotional showcase for natural commodities and a place to study nature for its own sake. Specimens were displayed for the free enjoyment of people from all walks of life. As the state's scientific authority, the museum gave visitors the prevailing scientific view of nature.

In the twentieth century the State Museum's collections grew to more than a million specimens spanning the geology, paleontology, and zoology of the Southeast. Today researchers use the collections to construct a model of nature that describes the biodiversity and biogeography of the region, and the biological connections beyond its borders. This book examines some of the collections research that has helped to construct the model of nature for our time.

Today's students will contribute to their generation's collective understanding of nature, as Lawson did 300 years ago. New concerns and a wider field of knowledge will inevitably lead to changes in their perceptions of nature. What will remain constant is the desire to know more about nature in all its diversity, and to do this these students will surely tap the knowledge preserved in the collections of North Carolina's State Museum of Natural Sciences.

The curators at the North Carolina State Museum of Natural Sciences made this book possible. Many curators, past and present, spent their careers developing the collections at the museum. They provided ideas for the focus of each chapter of this book and gave me detailed accounts of their research. It seems to be in the blood of museum curators to give freely of their time and knowledge to help the public enjoy and appreciate the natural world. My privilege and delight has been to encounter the passion they have for natural history.

The original idea for this book came from museum director Betsy Bennett. Her dogged determination and enthusiasm truly brought the book into being. The devotion to the traditions of the museum shown by longtime staff member Eloise Potter served as an inspiration and guiding light. David Perry of the University of North Carolina Press refined and enlarged the theme. Museum director of research Stephen Busack spent many hours working to improve the manuscript and ensure its accuracy in every discipline. His help was invaluable. Museum staff members Karen Kemp and Nancy Walters offered important suggestions, as did Friends of the Museum staff Pat Gruska and member Jeanne Visnaw. Most helpful in finding resources were Janet Edgerton, Margaret Cotrufo, and Ken Draves of the museum's H. H. Brimley Library. Special thanks are due to Rosamond Purcell, for her encouragement in the writing of this book, and for sharing her insight on the relationship between science and culture. Much appreciated help with the assembling of

Fresh off the boat from England in 1700, young John Lawson immersed himself in the wild country of North Carolina. The landscape, he wrote in *A New Voyage to Carolina*, is "adorn'd by Nature with a pleasant Verdure, and beautiful Flowers, frequent in no other Places . . . large and spacious Rivers, pleasant Savanna's, and fine Meadows . . . proper Habitations for the Sweet-singing Birds." Soon Lawson was sending specimens of the verdure—the rare flowers, the minerals, and the animals—of this exotic land to eager collectors in Europe. He planned "to make a strict collection" of North Carolina's beasts, birds, fishes, insects, fossils, minerals, plants, and soils. For the next two centuries, museums in Europe and America would seek out North Carolina's natural treasures as prize specimens in their collections.

By the nineteenth century, natural history collections had moved from the private domain of wealthy men and learned societies to the public sector of the natural history museum. Museums bore the cultural role of public interpreter of nature, a role now shared with television, magazines, popular films, and the Internet.

This book looks at North Carolina's natural landscape through the experience of the North Carolina State Museum of Natural Sciences, whose collections and interpretive exhibits reflect the relationship between society and nature. The emphasis of the museum's collections changed as people changed the way they valued the natural resources of their state. When the museum was founded in 1879, its collections of flora and fauna shared space

Summer tanagers, *Piranga rubra*. Courtesy of Rosamond Purcell.

A Long Look at Nature

Collecting Nature

For more than a hundred years, curators, naturalists, and everyday citizens have deposited treasures from nature—rocks and sharks' teeth, albino squirrels and meteorites, palmetto wood and warbler eggs—in the collections of the North Carolina State Museum of Natural Sciences. Specimens come from well-documented field biology surveys and from people who study nature in their spare time—from explorers atop Mount Mitchell to fishermen off the continental shelf.

Ever since the first Englishmen surveyed the flora and fauna of Roanoke Island in 1584, natural history collections have helped people understand the complexity of nature in North Carolina. Over time, these collections reflect what we know and value about nature in each generation. Cultural attitudes and economic interests influence the growth of museum collections. In turn, collections provide the reference material scientists need to model the natural world in all its diversity.

Western science looks for order in this diversity. The array of specimens in a collection suggests patterns to the scientist, who searches for relationship among the parts of the whole. The quest for order is at root a cultural phenomenon; Europeans sought to understand the bewildering biological diversity and geology of the New World by collecting it. Naturalists were on the front lines of the European conquest of the Americas. Nature was a fount of potentially profitable resources, and naturalists were encouraged by their benefactors to bring home every specimen possible. Fame and fortune,

"Geology and the different Branches of Natural History . . . change the whole face of nature" (Elisha Mitchell, *Geology of North Carolina*, 1842).

Tympanic bulla (earbones) from fossil whales. Courtesy of Rosamond Purcell.

along with delight in learning and sincere appreciation of nature, continue to influence the direction of research in the natural sciences, and affect our concepts of the natural world.

Scientific theory results from repeated, quantifiable observation that provides no-nonsense data. Constructed from these data, the collections-based model of nature provides the material for the sorts of comparisons that society holds dear. The model (as interpreted in contemporary exhibits at the State Museum) reveals that North Carolina has

- the highest mountain east of the Mississippi River—Mount Mitchell at 6,684 feet
- the greatest density of salamander species in the world—58 species
- the greatest diversity of carnivorous plants in the world (tied with Florida)—31 species in 4 families
- the oldest stand of living trees in the East—1,700-year-old bald cypress trees on the Black River
- the greatest diversity of fungi in the continental United States—more than 8,000 species.

European scientists of the sixteenth century also counted and sorted, verified observations, and shared their findings through the written word. Their method of research constructed a model of the New World, a model that was used to attract settlers. The birds of Roanoke Island were duly surveyed and documented by the Roanoke expedition's scientist, Thomas Hariot, in 1585. Unlike many European naturalists who were to follow, Hariot took the trouble to learn the Indian names of birds he observed. "Of all sorts of fowl, I have names in the country language, of fourscore and six . . . of several strange sorts of water fowl eight, and seventeen kinds more of land fowl . . . upon further discovery with their strange beasts, fish, trees, plants and herbs, they shall be published."

In contrast to Hariot and other naturalists, Native Americans possessed an intimate knowledge of the natural world passed from one generation to the next through oral tradition. Threads of this tradition remain among traditional Cherokee of the southern Appalachians, whose arts and religion are deeply connected to nature. Traditional Cherokee understood nature through a personal relationship with the natural world, in which knowledge was gained by revelation. In the 1990s a Cherokee elder, the Reverend Robert Bushyhead, recorded his relatives' teachings about plants and animals, forests and stars. His aunt taught him to find medicinal herbs by allowing the right plant to reveal itself. His father told him how the Cherokee learned the ways of animals: "A person was treated at a very early age by the medicine man to become a good hunter, [with] the ability to change himself into another figure spiritually. He could change himself into a form of a deer, spiritually . . . and he could go out there among the deer and tell which way they were traveling and how fast they were traveling."

(opposite)
Scarlet tanagers, *Piranga olivacea*. Collections preserve specimens—pinned insects, bottled fishes, trays of mollusk shells, drawers of minerals, rows of shrew skulls, tissues, eggshells, feathers, bones, nests, and skins—along with photographs, video and audio recordings, field notes, and meticulous drawings. Courtesy of Rosamond Purcell.

Hariot was confident that Native Americans would adopt a European approach to nature. "Upon due consideration [they] shall finde our manner of knowledges . . . to exceed theirs in perfection." However, the English settlers at Roanoke soon failed in the new land, perhaps in part because they were oblivious to the ways of nature that native people had come to understand over centuries of successful habitation in the same area.

Vivid accounts of New World flora and fauna enticed wealthy European collectors to covet exotic specimens from the Americas. Over the course of centuries, intrepid naturalists shipped many boatloads of stuffed, pressed, pickled, and live plants and animals across the Atlantic to fill the "cabinets of curiosity" in noble homes and the collections of early museums. John Lawson, the eighteenth-century explorer, collected specimens in North Carolina for natural scientists and wealthy men in London, shipping everything from plants and flowers pressed between sheets of paper to snakes, lizards, and small birds bottled in a homebrew of "aloes, myrrh, allom & tobacco steept in rum."

In his 1709 account, *A New Voyage to Carolina*, Lawson gave Europeans a tantalizing glimpse of the region's natural treasures. A keen observer, Lawson explored the uncharted Piedmont north to the Yadkin River valley and then trekked eastward to the English settlements on the "Pampticough River" near present-day Washington. North Carolina's great diversity of unspoiled habitats impressed the young naturalist, who found lands "here barren of Pine, but affording Pitch, Tar, and Masts; there vastly rich, especially on the Freshes of the Rivers, one part bearing great Timbers, others being Savanna's or natural Meads, where no Trees grow for several Miles, adorn'd by Nature with a pleasant Verdure, and beautiful Flowers, frequent in no other Places."

The number of species known to Western science skyrocketed in the seventeenth and eighteenth centuries. Europeans sought an all-encompassing classification system, or taxonomy, that would place the world's animals and plants in logical relation to each other. They devised schemes that ordered specimens by size or shape, or grouped them according to habitat or behavior.

Most classifications collapsed when the diversity of nature failed to fit into neat categories. In one early system, all limbless creatures were grouped together, including snakes, worms, and slugs. Lawson placed turtles and snakes, alligators and lizards "among the Insects, because they lay eggs, and I did not know well where to put them." Like others of his day, Lawson listed whales and dolphins among the fishes, and grouped as shellfish "crabs . . . craw-fish . . . tortois and terebin [turtles] . . . Finger-Fish [starfish] . . . and Oysters great and small."

In Sweden, Carl Linnaeus amassed a collection of 40,000 plant and animal specimens from Europe and abroad. He sorted plants by their sexual characteristics into 24 classes, which he subdivided into orders, genera, and

A Native American of coastal North Carolina. Native American names for scores of birds from the coast of North Carolina were recorded by Thomas Hariot in the 1580s. Native names for plants and animals were eventually lost as Europeans named and described species according to Western scientific taxonomy. Drawing by John White, 1584.

Smooth dogfish, *Mustelus canis* (*left*); "Beasts of Carolina" (*above*). New World naturalists sent exotic specimens from the Americas to eager European collectors. John Lawson trafficked in plants and animals from North Carolina in the early 1700s. Dogfish photograph courtesy of Rosamond Purcell; "Beasts" from Lawson, *A New Voyage to Carolina*, 1709.

species. To each species he assigned two names, a convention that continues today unchanged from 1758, when the tenth edition of his *Systema Naturae* was published. A comprehensive classification of 12,000 species, Linnaeus's taxonomy ordered the world's plants and animals then known to science.

Linnaeus's basic system, still in use, is typically used to organize specimens in a collection by taxa: species, genus, family, order, class, phylum, kingdom. The closest relatives are in the same species, closely related species are in the same genus, and so on up the organizational ladder. Linnaeus divided living creatures into two kingdoms—plant and animal. Still a matter of debate among taxonomists, the number of kingdoms now varies from five to thirty or more, including kingdoms of fungi, slime molds, viruses, and bacteria. In the animal kingdom, invertebrates, or animals without backbones — a huge group that includes insects, clams, corals, and jellyfish — account for all but one of the thirty or so phyla now recognized. The phylum Chordata includes all vertebrates: fishes, reptiles, amphibians, birds, and mammals.

Linnaeus gained his knowledge of the world's species through a voluminous correspondence with international naturalists. American botanist John Bartram dazzled the Swedish taxonomist with a description of the amazing Venus flytrap, found only in coastal Carolina savannas. A botanizing colonial governor of North Carolina, Arthur Dobbs, was probably the first to document the carnivorous species, in a letter to English collector Peter Collinson: "We have a kind of Catch Fly Sensitive which closes upon anything that touches it. It grows in this latitude 34 but not in 35°."

Bartram's naturalist son William, supported by Collinson and a network of naturalist-collectors on both sides of the Atlantic, described "those sportive vegetables" of North Carolina in his 1791 *Travels*: "Astonishing production! See the incarnate lobes expanding, how gay and ludicrous they appear! Ready on the spring to intrap incautious deluded insects, what artifice! There behold one of the leaves just closed upon a struggling fly, another has got a worm, its hold is sure, its prey can never escape — carnivorous vegetable!" William Bartram documented much of the flora of the Southeast, using Linnaeus's works as reference. The late publication of his *Travels* deprived him of credit for many of the new species he named.

Like many naturalists who were to follow, Bartram was awestruck by the extraordinary geology of the southern Appalachians and the diversity of the natural communities he encountered. His description of mountain cove habitat is still inviting: "We . . . mounted their steep ascents, rising gradually by ridges or steps one above another, frequently crossing narrow, fertile dales as we ascended; the air feels cool and animating, being charged with the fragrant breath of the mountain beauties, the blooming mountain cluster Rose, blushing Rhododendron and fair Lilly of the valley."

Bartram recognized that many species of plants with "northern affinities" grow in the mountains of North Carolina, and noted the effect of high-elevation climate on the plants of the region: "I began again to ascend the Jore [Nantahala] mountains, which I at length accomplished, and rested on the most elevated peak; from whence I beheld with rapture and astonishment, a sublimely awful scene of power and magnificence, a world of mountains piled upon mountains . . . at a distance surrounded with high forests, I was on this elevated region sensible of an alteration in the air, from warm to cold, and found that vegetation was here greatly behind, in plants of the same kind in the country below."

A passion for natural history collecting reached new heights in the nineteenth century, fueled by discoveries in geology and paleontology, and by the burst of newly described species of modern plants and animals. Influenced by a British movement called natural theology, many collectors believed that

such diversity was evidence of a grand design. America's leading scientist, Louis Agassiz, wrote, "All these facts in their natural connection proclaim aloud the One God . . . Natural History must in good time become the analysis of the thoughts of the Creator of the Universe." Observations of the natural world were clues to the nature of God; a path led "from Nature, to Nature's God." Clergymen often doubled as naturalists, including prominent North Carolina botanist the Reverend Moses Ashley Curtis and geologist Elisha Mitchell. Mitchell captured the intellectual excitement of this era of changing paradigms, writing in 1842, "Geology and the different Branches of Natural History change the whole face of nature."

If theology inspired collecting, economics funded it. Both botany and geology held commercial potential for North Carolina's mining and forestry enterprises. A gold rush followed the discovery of a 17-pound gold specimen in Cabarrus County in 1799. Convinced of the benefit of exploring the untapped mineral wealth of the state, the North Carolina General Assembly funded one of the first state geological surveys in the nation. Denison Olmsted, a divinity student and later geology professor at the University of North Carolina, rode horseback from Cape Lookout to the Smoky Mountains, gathering specimens and data for the initial survey. His mineral specimens, along with those of his successor at the university, Elisha Mitchell, may have been part of an informal museum in the old state capitol as early as the 1820s.

Venus flytrap, *Dionaea muscipula*. The diversity of species in the New World challenged the natural history classifications devised by European naturalists. Taxonomist Carl Linnaeus was fascinated by the carnivorous behavior of the Venus flytrap of coastal North Carolina. Courtesy of Jim Page.

North Carolina's first State Geologist, Ebenezer Emmons, kept a cabinet of minerals, including minerals, ores, and fossils, in his office in the state capitol. This room, recreated in its original third floor space at the top of winding stone stairs, was a precursor of the State Museum. Courtesy of Charles Jones.

North Carolina hired its first state geologist, Ebenezer Emmons, in 1851. To keep an eye on his progress with the North Carolina Geological, Mineralogical, Botanical and Agricultural Survey, the legislature required Emmons to maintain a collection called the Cabinet of Minerals in the state capitol. At the close of the Civil War, his rocks, minerals, and fossils were pilfered by General Sherman's Union troops, despite Governor Zebulon Vance's request that the capitol with its "library and museum . . . be spared."

For help in surveying the state's natural history, Emmons turned to Curtis, an Episcopal minister who studied the animals and plants of the Coastal Plain, Piedmont, and mountains while pursuing his clerical assignments. He knew the flora and fauna of North Carolina as well as anyone, and was the leading authority in the nation on mushrooms and other fungi.

Curtis's devotion to taxonomy was unrivaled in the state. In an 1834 letter to his fiancée, he tried to explain the appeal of his unpaid labors:

> At one time you might find me in the midst of 20,000 or 30,000 volumes, poring over tomes ancient and modern, folios and octodecimos, collating from Linneus (sic) down to Eaton to settle the obscurity that involves some of your common weeds. At another time you would see me, and make wry faces too, in the midst of a cabinet of minerals, monkeys, birds, fishes, bugs, shells, skeletons and snakes, surrounded by the beautiful and frightful of nature, the attractive and repulsive in life, but all interesting and instructive, in their economy, habits and complicated mechanism. It is a wonder to me that Nature in all its features is not admired, from that which is 'awfully great' to that which is 'elegantly little.'

The state survey published Curtis's *The Woody Plants of the State* in 1860, and later issued a companion botanical catalog that recorded more than 4,800 species of North Carolina plants. *Woody Plants* was a practical guide intended to assist the layman in knowing and using the state's trees. Curtis set the direction for many State Museum publications that were to follow—collections-based works that helped ordinary people to enjoy nature. He intended to "make this essay of popular service, and as intelligible as possible to those who know nothing of systems and would not . . . master a scientific treatise." Nevertheless, a later edition claimed that *Woody Plants* raised North Carolina's stature in the scientific community, disseminating "knowledge of her singular botanical wealth, which had engaged the interest . . . of the most famous European and American Botanists for nearly one hundred years."

Curtis and Emmons realized that the unusually high diversity of the state's plants and animals was tied to its unique geographical position. About midway on the continental shoreline, North Carolina is a biological border state where the ranges of many southern and northern species of plants and animals overlap. Emmons pushed for legislative support of natural history research based on the state's biological significance: "The position of this state is such that it forms the north and south limits of many interesting products of Natural History, belonging both to the vegetable and animal kingdoms; and it has been regarded an important work to fix definitely the true north and south boundaries of species belonging to these kingdoms."

But once the practical botanical manual was published, economic concerns in the Reconstruction-era legislature ruled against further collections research. Despite entreaties from the nation's top scientists—botanist Asa Gray wrote that Curtis's plant collection was "invaluable" to the state, and Louis Agassiz maintained that Curtis's collection was a part of the progress of civilization—the state failed to support zoological surveys. No funds were available to publish Curtis's volume on the amphibians, reptiles, birds, and mammals of North Carolina. As no zoological museum then existed in the

Eastern box turtles, *Terrapene carolina* (*bottom*). The Reverend Moses Ashley Curtis (*below*) cataloged the state's plants and animals for the North Carolina Geological Survey. His 1860 volume, *The Woody Plants of the State*, brought attention to North Carolina's timber resources, but the state failed to publish his survey of amphibians, reptiles, birds, and mammals. Turtles photograph courtesy of Rosamond Purcell; Curtis photograph courtesy of the North Carolina Division of Archives and History.

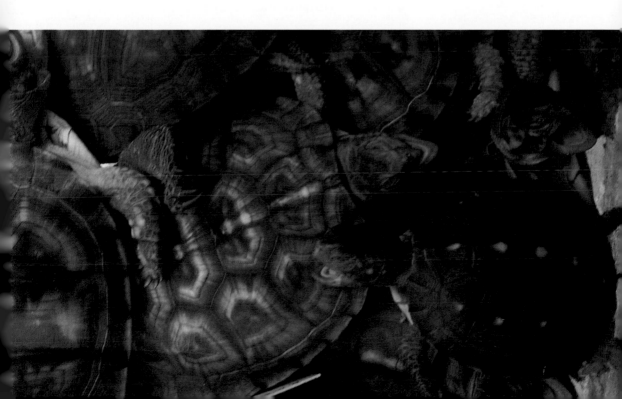

state, he sent his specimens of birds, reptiles, and mammals to the Smithsonian Institution for preservation. His large personal collection of bird specimens, including one of the few specimens collected in North Carolina of the now-extinct passenger pigeon, deteriorated after his death in 1872.

Other nineteenth-century naturalists safeguarded North Carolina specimens in northern museums. Dr. Elliott Coues deposited in the Smithsonian study skins of the birds of Fort Macon in coastal Carteret County. Dr. William Brewster collected mountain birds for the Museum of Comparative Zoology at Harvard. Albert Bickmore collected North Carolina marine mollusks, also for the Museum of Comparative Zoology.

Postwar American museums gained momentum as Northern industrialists poured money into public institutions. The field of natural history gained prominence in the 1860s and 1870s as scientists and the public debated the evolutionary theory of the century's foremost collector of nature, Charles Darwin. In North Carolina desire grew among scientists, state agencies, and the public for a museum for the state's natural resources.

A museum would serve as a signal of progress to the Northern victors, whose regard for Southern interest in natural history surfaced in a fanciful story published by the American Museum of Natural History. The story setting is the beach at Fort Macon during Union occupation. Union soldier Albert Bickmore, future founder of the American Museum, is collecting mollusks. Two Confederates train their rifles on him. "'It's only him again, that crazy Yankee shell-hunter.' The other Confederate grunts in disgust and drawls, "Third time this week we'uns hev had 'im in our sights. Wondah what he wants with them ole clam shells anyhow?"

In fact, visitors to the postwar state capital, Raleigh, could visit two collections ably curated by ex-Confederates. State Geologist Washington Carunthers Kerr displayed the state's collection of rocks, minerals, fossils, and woods, and Commissioner of Agriculture Colonel Leonidas LaFayette Polk kept a "Patron's State Museum" of agricultural products. Combining the two collections in a frugal piece of legislation in 1879, the General Assembly created the North Carolina State Museum "to illustrate the agricultural and other resources and the natural history of the State." Visitors could enjoy geology, agriculture, and a small but growing collection of botanical and zoological specimens in one location, under one name.

Many of the early specimens in the State Museum were part of promotional exhibits sent to world trade fairs in the United States and abroad. Kerr was probably the first state official who understood the potential payoff in exhibiting the state's resources to recruit out-of-state entrepreneurs. At his own expense he transported rocks and minerals from the state collection to the International Exposition in Vienna in 1873 and the Philadelphia Exposition of 1876. Kerr's efforts led the state to promote its resources at later fairs—Atlanta in 1881, Boston in 1883, New Orleans in 1885, Chicago in 1893, Paris in 1900, St. Louis in 1904.

Kerr hoped that North Carolinians would change their ingrained perceptions about the natural world. He wrote, "The people of the State have been accustomed to regard and to treat the forests as a natural enemy, to be extirpated, like their aboriginal denizens, human and feral, by all means and at any cost." With the new science of forestry, "all that has been changed." The new museum exhibits offered nineteenth-century citizens an alternative view of nature: that is, a resource to exploit, not a foe to subdue.

Gleaming wood sections of curly poplar and wild cherry stood near 12-inch spheres of polished granite, marble, and leopardite promoting the state's famed building stone. Porcelain plates made with mountain kaolin and building bricks of Piedmont clays were displayed along with gold, silver, and copper ores. Agricultural products from tobacco to cotton to silk represented farm life in most of the counties of North Carolina.

By 1897 the collections filled three exhibit rooms in the Agriculture Building across from the state capitol. The museum catalog listed 169 pages of mineral specimens and building stones; 50 pages of timber specimens, native woods, and medicinal plants; and 35 pages of agricultural products. "Mounted Specimens in Natural History Room" filled a mere 20 pages, including 79 lots of bird eggs. The names of museum exhibition halls reflected the concerns of commerce more than investigations in pure natural history—economic and scientific botany; economic entomology, forestry (including furniture products), history, and commercial fisheries. A lecture room with stereopticon capability was added in 1902 for presenting talks on farming and tourism. State pride in the growing museum found voice in the press. Journalist Fred A. Olds predicted, "The museum will be the greatest in this country outside Washington."

Promotional exhibits helped establish North Carolina markets for timber, minerals, and agricultural products. Interest in the exhibit of wood products

led state officials to speculate that "the time is at hand when the forests of North Carolina, if properly worked, will yield larger income than all her beds of gold." Collected by foresters W. W. Ashe and Gifford Pinchot in the 1890s, cross-sections from 130 native trees formed "a most useful commercial exhibit of forest growth." As forestry, mining, fisheries, and agriculture became the focus of specialized research sections in state agencies and universities, collections of plants, minerals, economic insects, commercial fishes, and shellfish were established at these institutions. The State Museum in the Agriculture Building in Raleigh remained the public showcase for the impressive tree sections and mineral specimens, but active collections of insects and plants were no longer kept there.

Scientific collections were the essence of European natural history museums, but they often played second fiddle to public educational exhibitions in the newly founded American museums of the 1880s and 1890s. Sponsored by wealthy industrialists, the natural history museums of New York City, Pittsburgh, and Chicago produced spectacular exhibits of dinosaur skeletons, wildlife of the American West, African mammals, and Asian birds, for the education and enjoyment of the public. Because North Carolina's wealthy families tended to endow colleges, hospitals, and art collections instead of natural history collections, the State Museum relied heavily on state funding.

Ensconced in the North Carolina Department of Agriculture, the museum veered between two pathways—educating and entertaining the public through exhibits and programs, and maintaining collections that served the needs of commerce, agriculture, and the natural sciences.

These paths were mirrored in the parallel careers of two English brothers, Herbert Hutchinson Brimley and Clement Samuel Brimley, who served North Carolina in separate capacities for nearly 60 years. One of the two, H. H., was a hunter-naturalist who pioneered interpretive exhibitions and programs at the State Museum; C. S. was a scientist-collector who painstakingly built the zoological collections that informed his brother's public offerings.

The brothers grew up in the collecting tradition of middle-class England, but times were tough and the Brimley family considered immigration to Australia or Canada. A chance meeting with an eager recruiter from the Department of Agriculture's Division of Immigration and Statistics convinced them that North Carolina held great promise for hard-working, genteel folk like themselves. When the Brimley family arrived in Raleigh in 1880, the brothers found new frontiers for natural history investigation, especially in the bird life of Wake County. In the tradition of John Lawson and the Bartrams, the enterprising brothers tried their hand at "collecting bird skins and eggs for wealthy men in the big cities, who vied with each other over the comparative magnitude of their collections," and were soon publishing papers in ornithological journals.

The Department of Agriculture contracted the elder brother, H. H., to create exhibits of game fishes and waterfowl that would attract well-heeled sportsmen to the state. A robust outdoorsman, H. H. played up the romance of the hunt in his exhibits, following a trend in natural history exhibitry both here and in Europe. His displays proved to be prize winners. In 1895 the department hired H. H. as the first full-time curator of the State Museum. Later he became the museum's first director.

Brimley readily embraced the American ideal of the museum as a vehicle for popular education, and applied himself to the distinctive promotional role of North Carolina's museum. He wrote, "A great good is being done by this institution both as an advertisement of the State's resources to visitors from beyond our borders and by its educational effect on North Carolina citizens."

But tension arose when the 1913 Board of Agriculture warned Brimley that his exhibits were wandering too far from a focus on farming. Brimley's passion was for natural history, especially the fishes, birds, reptiles, and whales. He had researched accounts for the scholarly work *The Fishes of North Carolina* and was in the midst of co-authoring the first bird book for North Carolina. Nature study and conservation movements were gaining ground in the state, and Brimley felt justified in nudging exhibits and collections closer to pure natural sciences. He gently reminded the Board of the 1879 state statute requiring the State Museum to illustrate the natural history of the state, and

Mounted specimens of birds and mammals, a leatherback turtle, and pinned insects (*below*) were prepared for display in the State Museum by curator H. H. Brimley, shown here (*right*) in 1901. North Carolina State Museum Archives.

added that "a long and close study of our visitors . . . proves very conclusively to me that a large majority prefer to observe and study objects of natural history above all other . . . exhibits."

Natural history gradually won the battle. Over the next 50 years the museum's exhibits reflected the interest in natural history shared by H. H. Brimley and the public, an interest often served by large, impressive specimens. In 1928 Brimley reported, "The big bull Sperm Whale, fifty-five feet in length, is the one outstanding specimen, both as to size and as to compara-

C. S. Brimley (*left*), a self-taught scientist, curated the zoological collections of insects, amphibians, reptiles, birds, and mammals that informed the museum exhibitry created by his brother, H. H. Brimley (*right*). North Carolina State Museum Archives.

tive rarity, that has ever come to the Museum. There may also be mentioned the large Ocean Sunfish, from Swansboro, estimated to weigh between 1200 and 1300 pounds; an eleven and a half foot alligator, a sail fish, an octopus with a five-foot spread of arms; a sixteen-foot Thresher Shark, from Wilmington; a very large Red Drum from Ocracoke, a yellow Raccoon, a pure Albino Opossum, a large specimen of Black Bear . . . the large cypress slab from Lenoir County . . . and a number of specimens of rare birds." A blend of boosterism and thoughtful science, the exhibits at the State Museum echoed the work of early English naturalists like John Lawson, whose passionate narratives mixed promotional hype with genuine awe at the diversity and beauty of nature in North Carolina.

C. S. Brimley worked alongside his brother, but collected for science, not for exhibition. At the age of 62 he modestly claimed that his "main interest for many years zoologically has been to gain and disseminate knowledge about the fauna of North Carolina, both vertebrates and invertebrates, with especial regard to Herpetology and Entomology." His legacy to the natural sciences includes more than 200 zoological papers, the landmark books *Insects of North Carolina* and *Birds of North Carolina*, and descriptions of new species of amphibians and insects. His detailed field records supported published surveys on the amphibians and reptiles of North Carolina, the mammals of North Carolina, and a partial account of fishes of North Carolina.

A self-taught scientist, C. S. initiated and indexed the zoological collections of the State Museum. His legendary concern for the state's herpetology collection has been handed down from one curator to the next through four generations. Among the many who have studied the natural sciences in North Carolina, C. S. Brimley comes closest to being the consummate naturalist foreseen by John Lawson in 1709: "when another Age is come, the Ingenious then in being . . . [will create] a complete Natural History [of North Carolina] ... no easie Undertaking to any Author that writes truly and compendiously."

The Brimley brothers died within three months of each other in 1946, having shaped the state's collections and public perceptions through six decades of great changes. They arrived in North Carolina when gray wolves and panthers still roamed the woods, and documented nature until the dawn of the atomic age. Each brother in his own way gave North Carolinians a broader understanding of the natural world, and both found in their adopted state and its museum a place to indulge their passion for nature.

Without the guiding presence of the Brimleys, the State Museum maintained its status quo in post-World War II America. Public attention riveted on the space race and physical sciences. Advances in cellular biology and biochemistry energized the scientific community, which sometimes dismissed systematics (collections-based studies) as a purely descriptive science devoid of theoretical interest—equivalent to stamp collecting. Universities unloaded their natural history collections on research museums. The National Science Foundation often sponsored these transfers, affirming the scientific value of these orphan collections. The State Museum's holdings, enriched by collections from universities and individuals, outgrew the confines of the museum building and were tucked into nooks and crannies of basements, lab buildings, and even a former ballroom.

Individuals remained devoted to the collections. Bill Palmer headed the museum's herpetology collection for 36 years, amassing a collection of 120,000 southeastern reptile and amphibian specimens that ranks among the finest in the world for regional collections. Vince Schneider curated the Museum's fossil collection for 16 years as a volunteer, working elsewhere as an environmental biologist. Curiosity about multi-legged locomotion led Rowland Shelley to assemble a millipede collection at the State Museum that ranks internationally. Biologists around the world use it to sort out this class of poorly understood but ecologically important animals.

(*opposite*)
Barking treefrogs, *Hyla gratiosa*. Natural history collections from some universities and agencies were donated to museums when cell biology and biochemistry became the focus of biological research after World War II. Courtesy of Rosamond Purcell.

Concern for endangered species, ecological threats, and environmental quality gathered momentum in the 1970s. New interest arose in the diversity of nature, and curators were tapped to conduct species inventories of development sites, to monitor populations of threatened species, and to provide collections data for environmental studies. In one case in the 1980s, scientists recorded high mercury levels in Lake Waccamaw. Was the mercury coming from pollutants in groundwater? A test of preserved Waccamaw fish specimens from the 1960s showed comparable levels of mercury—indicating no increase in groundwater mercury in 20 years.

In the 1990s, the State Museum's research collection grew to more than a million specimens in eight divisions spanning the geology, paleontology, invertebrates, and vertebrates of the Southeast. Today, curators pursue research in systematics (the study of the diversity of organisms and their evolutionary relationships), biogeography, and geological processes. Such studies contribute to a scientific model of the natural world that describes which organisms occur, how organisms interact with their environment, and how the natural word changes over time.

Identification of organisms is still the curator's bread and butter, and is essential information for biologists studying ecology, evolution, medicine, and environmental sciences. To make an accurate identification, a curator compares a specimen with the appropriate series in a collection. A series consists of several individuals of a species, including males and females of various ages, at each season of the year, from different locations. The series shows how individuals differ within a species and among similar species.

A curator who notices unique biological features in a group of like speci-

mens may make the case for naming a new species. Most biologists define a species as a population whose members can interbreed freely under natural conditions, although taxonomists use several attributes to describe a new species. Working in the collection, a curator may examine the physical traits, biochemical makeup, range of occurrence, and other factors that distinguish the group in question from similar species. Consensus among taxonomists on the validity of a new species is not easily won, and requires long hours reviewing the scientific literature and comparing series from various collections.

The lifework of curators and the delight of amateur naturalists, natural history collections are very much a product of culture. Early specimens in the State Museum collections reflected the increased importance of commercial fisheries and forest products and the director's interest in whaling on the Outer Banks. At the beginning of the twentieth century, the conservation movement—inspired in part by the decimation of North Carolina's ducks and shorebirds—resulted in expanded bird collections. Agricultural pests and venomous snakes sparked field surveys in the Depression and war years. Public concern leading to the Endangered Species Act of 1973 provided funds for curators to study little-known species like the gopher frog and the Rafinesque's big-eared bat, and to conduct faunal surveys in areas under development. In the 1980s, "global thinking" linked the bird collection with Partners in Flight, an international group that studies Neotropical birds, many of which migrate through North Carolina. Big acquisitions still nurture state pride: a 110-million-year-old dinosaur and 66-foot-long blue whale entered the State Museum's collections to public acclaim in the 1990s.

Spread wings of tropical birds (*facing page*). Knobbed whelks, *Busycon carica* (*above*). A natural history collection is a library of information. Specimens may yield valuable information for research many years after they are collected. Records tell when, where, and how each specimen was collected, and by whom. Courtesy of Rosamond Purcell.

Skeleton of True's beaked whale, *Mesoplodon mirus*. Courtesy of Rosamond Purcell.

Collections provide scholar and museum visitor, field biologist and backyard naturalist the means to explore the diversity of nature. As our knowledge expands and alters our perception of the natural world—changing "the whole face of nature," as Elisha Mitchell put it—our sense of responsibility has grown as well. In every collection, curators and others have contributed more than data to their generation's model of the natural world. They have voiced an affinity and concern for that world and the species that inhabit it, "from that which is awfully great to that which is elegantly little."

Rocks and Minerals

Rocks and minerals hold the clues to understanding the big picture in nature. Bedrock dictates the lay of the land, the type of soil, the availability of fresh water, and ultimately, the plants and animals in an area. Mount Mitchell, the highest peak in eastern North America, shares plant and animal species with Canadian forests. Subtropical animals and plants live on Jockey's Ridge, one of the highest sand dunes on the East Coast. With its landforms from high mountain to barrier island, North Carolina prompted eighteenth-century naturalist John Lawson to write, "One Part of this Country affords what the other is wholly a Stranger to."

Early naturalists noted the changes in plant and animal species, geologic features, and climate as they traveled from the tallest peaks in the west to the subtropical islands of the coast. They labeled the landscape just as they classified its plants, animals, and minerals. Four physiographic regions were obvious—the mountains, defined by the rugged Blue Ridge; the Piedmont with its rolling hills; the flat Coastal Plain; and the open ocean.

Along with extremes in landforms, North Carolina possesses an extraordinary diversity of minerals—making it a magnet for miners and geologists over the centuries. Gold drew Spanish explorers to the mountains in the 1540s. In 2000, North Carolina was number one in the nation in the mining of scrap mica, feldspar, and olivine. Piedmont clay deposits support the country's largest brick industry. Many specimens in the State Museum's col-

"It is through the discoveries in this department of science [geology] that we obtain a knowledge of the ancient history of the globe . . . no one will regret that he devoted a portion of his leisure hours to its study" (Emmons, *Report of the North Carolina Geological Survey*, 1858).

Chalcopyrite, mineral specimen. Courtesy of Rosamond Purcell.

lection of rocks and minerals are byproducts of the considerable economic activity involving the rocks of the state.

The first miners were pre-Columbian peoples. In the Mountain region, Native Americans left elaborate tunnels in ancient mica mines. Mitchell County miners in the nineteenth century cut into old shafts and tunnels that honeycombed the area. State Geologist Washington Caruthers Kerr visited the mines in the 1860s, reporting that "there are hundreds of old pits and connecting tunnels among the spurs and knobs and ridges of this rugged region; and there remains no doubt that mining was carried on here for ages, and in a very systematic and skillful way."

Native Americans near Morrow Mountain traded in rhyolite, a very hard volcanic rock used by many tribes for making stone tools. The Cherokee may have mined gold. Rumors of gold were sturdy enough to lure Spanish explorer Hernando de Soto to Cherokee country in 1540. Decades later, scientist Thomas Hariot of the 1585 English expedition to coastal North Carolina documented the use of copper ornaments by Native Americans. He also noted the economic potential of iron-rich rock further inland. At the turn of the eighteenth century, John Lawson reported that "iron-stone we have plenty of, both in the Low-Grounds and on the Hills." English ships exported the first shipments of Carolina iron in 1729.

More useful to the colonials were Carolina clays, which made "good Bricks and Tiles" according to Lawson, who listed "several sorts of useful Earths . . . in great plenty; Earths for the Potters Trade, and fine Sand for the Glass-makers." Famed English potter Josiah Wedgwood coveted a vein of kaolin, a residual clay used in fine porcelain, on Cherokee lands in Macon

Mineral specimens. Interest in the diverse mineral resources of North Carolina instigated the first natural sciences research funded by the state. Courtesy of Rosamond Purcell.

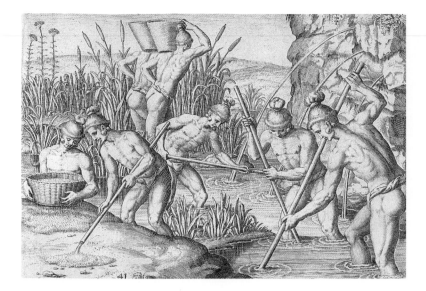

County. He arranged to ship five tons of the rare clay, mined by the Cherokee, to his pottery in England in 1767.

Clays from the volcanic slate beds of the Carolina Terrane and shales from the Triassic basins supplied new brick-making businesses as Europeans settled the region. By 1776 potteries dotted the Piedmont from the upper Cape Fear River valley to the Moravian settlements of Wachovia to the Catawba Valley. Renowned North Carolina pottery from native clays is still in great demand today. Generations of potters have mined local clay "ponds" for residual clay, derived from local rock, and sedimentary clays, formed from sediments deposited by streams. Residual clays tend to be freer of impurities and less plastic than sedimentary clays.

Botanist William Bartram described the clays of the Cape Fear River valley in the 1770s. Clays could be found in the "flat level land back from the rivers, where the clays or marle approach very near the surface, and the ridges of the sand hills, where the clays lie much deeper." Mineral springs gave naturalists a clue to the state's geology. Bartram used all five senses to observe the evidence:

> "Almost imperceptible veins or strata of fine micaceous particles, which drain or percolate a clear water, continually exuding, or trickling down, and form in little rills and diminutive cataracts . . . in some places, a portion of this clear water or transparent vapour, seems to coagulate on the edges of the veins and fissures, leaving a reddish curd or jelly-like substance sticking to them, which I should suppose indicates it to spring from a ferruginous [iron-rich] source, especially since it discovers a chalybeate scent and taste: in other places these fixtures shew evidently a crystallization of exceeding fine white salts, which have an aluminous or vitriolic scent: there are pyrites, marcasites, or sulphureous nodules, shining like brass of various sizes and forms."

At the turn of the nineteenth century, gold transformed North Carolina's economy and interest in geology surged. Gold mines may have operated in the state before the Revolutionary War, but 1799 marks the first documented discovery of gold in the state. Young Conrad Reed discovered a bright yellow rock in a creek on the family farm in Cabarrus County. He lugged the 17-pound specimen home, where it served as a doorstop for years. Conrad's father, John, finally sold the gold nugget for $3.50. When Reed learned that its true worth was around $3500, he set out to find more nuggets. His farm became the Reed Gold Mine. In a hundred years of operation, it produced more than 100 pounds of big nuggets and made John Reed a very wealthy man. Other farmers found gold in their creeks, and in 1825 a Stanly County farmer dug into a gold-bearing quartz vein near his creekside operation. His discovery ushered in underground, or lode, mining in the state.

Miners, engineers, and adventurers flocked to North Carolina, introducing steam engines and gold coinage. Gold Hill in Rowan County became a wide-open boomtown with 15 mines and 27 saloons. The U.S. Mint opened a branch in Charlotte, the center of the robust mining economy. Gold mining was second only to farming as a moneymaker in North Carolina, the nation's top gold producer at a million dollars a year before the 1849 California gold rush. Most North Carolina mines closed during the Civil War, though some re-opened and remained profitable until the early 1900s.

A fever for gold and minerals with newfound industrial uses provided the momentum for the first state-sponsored natural history collections. Denison Olmsted, an energetic young chemistry professor at the University of North Carolina, launched a study he called "Useful Minerals in North Carolina," and received funds from the state legislature for a statewide survey of rocks and minerals. Olmsted supplied the state legislature with a geologic map, the first in the country to be produced with state funds. His 1825 map outlined features well known today: the Coastal Plain, the Deep River Triassic basin, the Great Slate Formation, the Dan River Triassic basin, and the Raleigh antiform.

The western Piedmont boomtown culture of the gold miners made an impression on Olmsted. "A 'Gold Hunter,'" he wrote, is "one of an order of people that begin already to be accounted a distinct race." Convinced that gold deposits were limited to that area, Olmsted thought gold must have been deposited by strong, ancient currents that made them "in short, to be standing monuments to the Deluge" (Noah's flood). While Olmsted's reading of the rocks was in line with the worldview of his era, geology would move away from a strictly biblical interpretation in the next few years.

Elisha Mitchell took Olmsted's place at the University of North Carolina in 1825 and continued the state survey for a while. He thought Olmsted was wrong about the origin of the state's gold, claiming instead that gold was released from rock veins by weathering. One of his students, politician (and later Confederate general) Thomas L. Clingman, found platinum in the mountains, as well as diamond, mica, and corundum. An unfortunate dispute between Clingman and Mitchell regarding the highest point in North Carolina led Mitchell to scale the nearly inaccessible peaks of the Black Mountains several times. Mitchell fell to his death on the last climb, before his original ascent was confirmed as the first measurement of the highest point east of the Mississippi River. Mount Mitchell measures 6,684 feet high. Clingman is credited as the first to measure the tallest point in the Great Smokies, the 6,642-foot-high Clingman's Dome.

By the middle of the nineteenth century, the state's interest in geology was sufficient to fund a new post, the office of state geologist. Professor Ebenezer Emmons of the New York Geologic Survey undertook the work. To keep track of Emmons's survey progress, the state legislature required him to display at the state capitol a cabinet of minerals, which some referred to as "the state museum." As did Olmsted and Mitchell, Emmons collected fossils for the collection, including mosasaur, crocodile, and turtle fossils from the Cape Fear region.

Emmons found in the volcanic slate formation in Moore County a "soapstone" used by locals for fireplaces and gravestones. He knew it was really agalmatalite, later called pyrophyllite. Mines soon opened in the area. Moore and surrounding counties are now the country's largest producers of pyrophyllite, used in ceramics and as filler in manufactured products.

A native of Massachusetts, Emmons retired from his post on the eve of civil war, to be replaced by Harvard-trained Confederate soldier W. C. Kerr. Professor Kerr spent the war years developing mineral resources for the Confederate cause: saltpeter, sulfur, iron, and coal. He most likely was not

Ebenezer Emmons, (*top*) North Carolina's first state geologist, collected a cabinet of minerals, which was housed in the state capitol in the 1850s. Courtesy of the North Carolina Division of Archives and History.

The Boss (Rossie) lead mine in Davidson County (*above*) was one of three lead mines in the state in 1856. Here Ebenezer Emmons found "handsome cabinet specimens of galena . . . from a very distinct vein of quarts." From Emmons, *Geological Report of the Midland Counties of North Carolina*, 1856.

present on April 13, 1865, as General Sherman's troops pilfered North Carolina's cabinet of minerals.

Kerr remained state geologist for 20 years after the Civil War, traveling
thousands of miles on horseback in a depressed postwar landscape to survey
unmapped areas in central and western North Carolina. Believing that the
state's mineral resources were the key to economic recovery, he exhibited at
his own expense rocks and minerals from the state collection at the International Exposition in Vienna in 1873 and the Philadelphia Exposition of 1876.
Kerr's efforts encouraged the state to promote its resources at later fairs in
Atlanta, Boston, and New Orleans in the 1880s. He set the stage for the extensive mineral displays of the Victorian-era State Museum.

Rocks and minerals were the biggest attraction by far in the State Museum
at the turn of the twentieth century. The state was banking on its mineral resources, which fetched a total of $24 million in 1894 dollars according to
North Carolina and Its Resources, a state handbook. Gold, copper, and iron led
the pack. Corundum, mica, talc, pyrophyllite (still identified as Emmons's
agalmatalite), and gems brought revenue to the state, as did Mount Airy's extensive granite quarries. Polished spheres and columns of North Carolina
granites and marbles crowded the museum's Geology Room. The 12-inch

CLEAVELAND SPRINGS, SHELBY, N. C.

Names of mineral springs resorts indicated the mineral content of their waters. Barium Springs, Chalybeate Springs, Lincoln Lithia Springs, and White Sulphur Springs were among those that offered health cures to nineteenth-century tourists. Courtesy of the North Carolina Division of Archives and History.

spheres showcased mountain marbles, rare orbicular diorite, leopardite, monumental granite from the Piedmont, and pink granite from the Coastal Plain. A selection of earthenware and stoneware by Piedmont potters showed off the state's clays. Gold, copper, and iron ores represented the wealth of the state's mines, along with mica, talc, quartz, gneiss, coal, graphite, and slate. Museum collection records from 1897 list gemstones "both cut and in the rough, some of the specimens being particularly well-colored, of great brilliance, and valuable," including hiddenite, emerald, garnet, aquamarine, topaz, ruby, and amethyst.

Not all of the state's mineral resources were mined or quarried—some were drunk. Fifty bottles of assorted mineral waters were displayed in the State Museum. By the mid-1800s, mineral springs were a North Carolina tourist attraction. Sixteen mineral spring resorts were promoted in the state handbook of 1896: "Certainly most all parts of the State boast of some mineral spring whose waters bring health by assisting nature in restoring the afflicted." Well-heeled out-of-state visitors sought cures for bladder and kidney complaints at White Sulphur Springs or for nervous diseases at Lincoln Lithia Springs. Ellerbee Springs was known for its iron and sulfur content, "helpful for hayfever sufferers." Barium Springs, Bromine-arsenic Springs, Red Springs, and Seven Springs enjoyed a loyal clientele; waters from Panacea Springs were bottled and exported to adjoining states.

Rocks from space, namely meteorites, stirred the imagination of earthbound North Carolinians long before the Wright brothers flew across the dunes of North Carolina's Outer Banks. "[M]eteorites are the only known communication we have with other worlds" wrote the State Museum's first director, H. H. Brimley, "and they all come one way." General Clingman's meteorite collection from the mountains and Piedmont was described in Kerr's 1875 survey report. Kerr acquired an exceptional meteorite, found in a Rockingham County field before the Civil War, for the collection. Geologist

Quartz. The Geological Room in the State Museum displayed specimens of quartz, mica, talc, gneiss, coal, graphite, and slate along with building stones, precious gems, meteorites, and gold ores. Courtesy of Rosamond Purcell.

Friedrich A. Genth analyzed the chemical composition of the stone and pronounced it "undoubtedly one of the most interesting in existence." Another scientist deposited a sliver of this meteorite in the Jardin des Plantes of the Museum National d'Histoire Naturelle in Paris.

Every day, approximately ten meteorites fall on Earth, mostly in the oceans or polar icecaps. Those few that are seen to land make headlines. General Clingman published an account of a meteor that fell in August 1860 in "a bright glare of light . . . equal in size to the full moon." A Dr. King reported "a shower of meteorites" in Nash County on May 14, 1874. Witnesses heard "rumbling explosions as of fire arms in battle a few miles off, which continued for four minutes." State Supreme Court Justice J. J. Davis secured some of the fallen stones for the museum.

In 1934 many people in North Carolina witnessed a meteorite as it fell to Earth on a west-to-east path from Gastonia to Farmville. It split in two pieces as it encountered the atmosphere, accompanied by an explosion that was heard for 250 miles. One piece fell in a yard in Farmville; farmers discovered the second piece during spring plowing the following season. Museum director and geologist Harry Davis, an avid student of meteorites, acquired the specimen for the State Museum, where it is still on exhibit today.

Perhaps the most valuable meteorite in the museum's collection fell on a farm in Moore County in 1913. Around 5 P.M. on April 21, George Graves and two hired hands heard a roaring overhead. They looked up to see a red ball trailing a 15-foot tail of blue-black smoke. Sparks shot from the stone, which landed a few feet from one of the men. About as big as a box of tissues, the four-pound meteorite dug a hole two feet deep in the newly plowed earth. Graves sold the stone for $100 to the Smithsonian Institution 20 years later.

Eventually the State Museum traded a portion of its Farmville meteorite to acquire a chunk of the Moore County meteorite.

If Graves and his workers had missed the sight of the blue-tailed fireball, the meteorite would have easily been mistaken for a terrestrial rock, and its extraordinary parentage would remain unknown. While most meteorites are made of metallic iron or iron with uncommon minerals, the Moore County meteorite is made of basaltic minerals, similar to the composition of Moon rocks and rocks from the deep ocean basins of Earth. It is as old as the solar system, billions of years older than North Carolina's oldest rocks. The Moore County meteorite and others of its type probably came from a parent asteroid that was on its way to being Earth-like. These celestial stones reveal something of the early geologic processes that formed our planet.

Because the Moore County meteorite reflects an unusual light spectrum, scientists were able to determine that its probable parent asteroid is 4 Vesta. The brightest asteroid in our sky, 4 Vesta has remained intact since the beginnings of the solar system 4.5 billion years ago. Like early Earth, it was once molten, and its heavy metals sunk to a dense core, while lighter rocks floated upwards. But then a larger body plowed into it, and the planet-forming process halted in mid-stream. Photographs of 4 Vesta taken by the Hubble Space Telescope actually show a vast crater blasted in the side of the asteroid. Meteorites from this type of proto-planet are specimens straight from the early solar system. They take scientists back in time to investigate the composition and size of early planets, the timing of planet-building processes, and possible heat sources essential to the making of a planet.

In the twentieth century, economic interests continued to back most geological studies in the state. Gold, copper, and iron took a backseat to other minerals used in industry and the building trades. The office of state geologist evolved into the Division of Land Resources, including the North Carolina Geological Survey and Mining Commission. State agencies and universities developed their own research collections; meanwhile the State Museum collection expanded with specimens from played-out mines.

The Moore County meteorite may be the most valuable meteorite to fall within the state's borders. It probably came from the asteroid 4 Vesta. Courtesy of the North Carolina Division of Archives and History.

During World War II, a new exhibit displayed minerals needed in the war effort: bauxite, mica, olivine, and chromite. Museum director Harry Davis analyzed specimens for prospectors looking to supply the country's mineral needs. A traveling exhibit was sent across the state to encourage the wartime quest for deposits of bauxite, used in the production of aluminum. Tungsten was discovered in Vance County, where the Hamme Mine briefly became the world's largest producer of tungsten. In the Coastal Plain, phosphates were mined for fertilizers; the mountains produced feldspar for ceramics and glassware, and talc for paints and ceramics. By the twenty-first century, North Carolina was second only to Florida in phosphate mining, and Mitchell County provided more than half of the feldspar used in the United States each year.

For its value as a building stone, granite won the title of official rock of the state of North Carolina in 1979. Found throughout the Piedmont and Mountain regions, the state's granites are a byproduct of the continental plate collisions that built the Appalachians. Folding and faulting pushed deep rock layers toward the surface; erosion exposed the rock in mountain outcrops like Looking Glass Rock in Transylvania County and Stone Mountain in Wilkes County. The mile-long Mount Airy granite quarry is the largest open-face granite quarry in the world, producing a top-quality stone used for buildings, bridges, and statues, as well as crushed stone for construction.

The emerald was crowned as North Carolina's official precious stone in 1973. Except for a single emerald discovery in Utah, North Carolina is the only place in North America known to have emerald veins. The state's first documented emerald, an 1875 find in the Alexander County town of Stony Point, touched off a spell of emerald fever. The Emerald and Hiddenite Mining Company started operations in 1881. Fine-quality emeralds cropped up in the mountains near Spruce Pine in 1894. Emeralds from these locations were on display in the State Museum by 1897. Emeralds discovered in the western Piedmont near Shelby in 1909 prompted the Emerald Company of America to open the Old Plantation Mine, which produced 3,000 carats worth of emeralds.

The quartz veins of Alexander County have produced the largest emeralds. In 1999 prospector James Hill found the largest North American emerald to date. A rock hound as a boy, Hill grew up with stories of gem mines and talked to older miners about likely lodes in Alexander County. His investigations led to emerald-rich veins and a giant 69-carat emerald. The big crystal was cut into two gems—the 18.88-carat Carolina Queen, and the 7.85-carat Carolina Prince, which sold for $500,000 in 1999.

Emeralds owe their existence to the chance meeting of beryllium and chromium. High concentrations of these elements are not found together in the same rocks, but they may come in contact if molten igneous rock high in beryllium flows into a fissure of metamorphic country rock containing chromium; the resulting crystal formation between the molten rock and

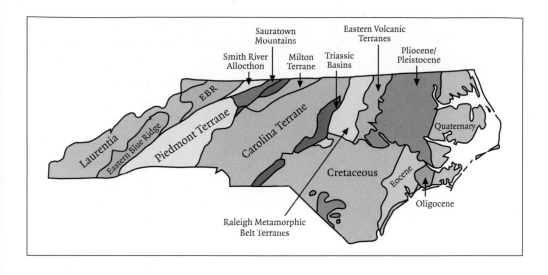

country rock is an emerald. A green version of the mineral beryl (made of beryllium, aluminum, and silicon), an emerald gets its color from chromium contamination. The Spruce Pine emeralds clearly illustrate the process—the white pegmatite stands in sharp contrast to surrounding black country rock. On the other hand, the background of the Alexander County emeralds is a puzzle. No likely rocks seem to have been present in the right place at the right time for emerald formation to occur.

Geologic map of North Carolina, simplified. Redrawn from original map by Chris Tacker.

North Carolina's unique geologic history is yet to be fully understood, but since the 1960s geologists have used a powerful theory to decode its secrets. Plate tectonics (from the Greek *tekton*, meaning builder, as in architect) explains the "building" of landforms, from mountains to island volcanoes, by the slow but sure movement of continental plates. These plates are made of light crust that floats on Earth's heavier mantle. North Carolina rock layers bear evidence of a history of collisions, rifts, and erosion that occurred over hundreds of millions of years as continental plates drifted over the globe.

The state's oldest rocks form the deep core of the Blue Ridge Mountains. The result of the collision of two continental plates more than a billion years ago, the rocks are all that is left of the mountains that were pushed up when the plates ground into each other. Rocks formed by this event, the Grenville Orogeny, are exposed today in scattered mountain outcrops from Canada to Alabama.

Around 734 million years ago, the early North American plate began to pull away from the larger landmass. Earth's crust stretched thin like taffy as the plates parted, forming rift basins that filled with sediments that produced the sedimentary conglomerate rocks at Grandfather Mountain.

The state's famous emeralds are part of the strongly deformed metamorphic rocks of the Piedmont Terrane. These rocks may have been part of a piece of the continent that tore away and reattached in a later collision. Some of these rocks contain olivine, a mineral formed as a result of collision, which is so stable at high temperatures it is used in steel foundries. Intense

Corundums, partly rimmed by margarite. Original label identifies the specimens from Buck Creek Mine, Clay County, which closed in the early 1900s. Some corundums were compressed into an unusual pinched bullet shape as chunks of rock were squeezed during metamorphism. Courtesy of Rosamond Purcell.

pressure and heat transformed the rocks in this region, producing garnet and corundum hard enough for use in abrasives. Rubies are red corundum in its clearest form; other colors of corundum are sapphires.

Much of North Carolina's gold originated in volcanic events that occurred 650 to 540 million years ago. As North America and Africa began to move toward each other, one oceanic plate forced the other downward, deep into the mantle of the Earth. The cold slab warmed up in the mantle and released water, lowering the melting point of surrounding mantle rocks. Magmas formed above the slab, rising to create volcanoes that grew into islands. Around these volcanoes, water superheated and dissolved silicon and gold. As the circulating water cooled, it precipitated gold and silicon in concentrated deposits in the volcanic rock. Around 450 to 480 million years ago, the string of volcanic islands, now known as the Carolina Terrane, was shoved on to the edge of the North American continent as Africa moved ever closer. The collision compressed the gold-bearing Terrane rocks at a relatively low temperature and pressure, creating the gold deposits that made nineteenth-century miners rich.

Africa began to collide with North America about 325 million years ago. Massive compression of the North American crust resulted in another mountain-building event, the Allegheny Orogeny, as the land buckled and folded along the eastern edge of the continent. In this final raising of the Appalachians, the Carolina and Piedmont Terranes piled up into mountains as high as the modern Andes. Some of the stress built up by the pushing plates was released in faults as the crust of one plate pushed up and over the opposing plate. Earthquakes were common along these faults, though fault zones are quiet today.

The African plate began to tear away from North America between 230 and 220 million years ago in the Triassic period. The crust stretched thin again, and new rift basins formed, filling with sediments eroding swiftly from the high mountains. Triassic rocks in the Deep River and Dan River basins contain fossils of phytosaurs that hunted for fish in rift lakes, in a landscape of tree ferns and cycads. Sedimentary shales, sandstone, and coal were deposited in the Piedmont rift region, as were the clays now used by brick-making businesses. Land currently east of Raleigh remained behind as Africa pulled away, leaving the young Atlantic Ocean in the widening gap between the two continents. Spreading continues today, as North America heads toward Asia at about the same rate as fingernails grow.

By the end of the Cretaceous period 65 million years ago, erosion had worn down the once towering Appalachians. Rivers carried mountain sediment downstream to the Coastal Plain, North Carolina's youngest geologic

(*top*)
Folding in rock layers at Chunky Gal Mountain, Macon County, was caused by the intense heat and pressure from the grinding of opposing continental plates hundreds of millions of years ago. Courtesy of Chris Tacker.

(*bottom*)
The sedimentary rock layers exposed in this Coastal Plain excavation at Aurora, Beaufort County, represent periods when ocean covered the land. Courtesy of Vince Schneider.

region. As sea level rose and fell with changes in climate, a stack of wedge-shaped layers of sedimentary rocks built up the flat Coastal Plain from east to west. Sands laid down in shallow Eocene seas formed the low-rolling Sandhills 55 to 38 million years ago. Distinctive Coastal Plain depressions called Carolina bays began to form between 200,000 and 100,000 years ago. Carolina bays were probably created by wind-driven currents, which carved the characteristic elliptical shapes in the loose sands of shallow lakes.

During the last ice age more than 18,000 years ago, sea level was lower and the continental shelf was above water. Rivers cut channels into the Coastal Plain rock, creating valleys. At the shoreline at the edge of the shelf, the bottom dropped steeply to meet the vast abyssal plains of the Atlantic Ocean. Near the end of the last ice age about 10,000 years ago, glaciers melted, sea level rose, and the ocean began to cover the shelf and its river valleys. Today these flooded ancient riverbanks provide hardbottom habitat for marine organisms adapted to living on hard surfaces. The barrier islands formed as rising waters encircled low ridges on the mainland. Still the first land to meet the fierce storms of the Atlantic, the thin string of islands and inlets is the state's most dynamic landform, changing shape with each season and storm.

North Carolina's rocks tell a story of earthquakes, lava flows, massive erosion, and the reshuffling of huge rock slabs resulting from continental collisions. Quartz and gold from deep in Earth's crust bear evidence of volcanic islands that crashed into the continent. The value of gold altered the course of human events in North Carolina, as did the presence of many other rocks and minerals. Economic interest initiated the first state funding of natural history collections.

North Carolina rock layers also hold a panorama of fossils from past ages, from the oldest multi-celled fossils in the Americas to the mastodons and ancient whales of the Pleistocene. Geologists and paleontologists continue to work side by side to unravel the long history of the natural world written in the rock.

Fossils

THREE

When life was mostly in the seas, a marine worm left its mark in the sediments of the Carolina Terrane near Durham. At 620 million years old, this trace fossil named *Vermiforma antiqua* is the oldest record of life in North Carolina. Originally below the equator, the rock that preserved the worm's shape was part of an ancient seabed. Seventy million years later, another early life form, the sea pen *Pteridinium carolinaensis*, left its impression in North Carolina rocks.

In addition to these early marine worms and sea pens, a succession of species preserved in stone over hundreds of millions of years prove that sago palms, great white sharks, giant crocodiles, dinosaurs, and mastodons lived in an area that is now called North Carolina. Geology dictates the kinds of fossils that survive within the state's borders. Big events that shaped the state—continental collision, mountain-building, Coastal Plain sedimentation—determined whether the fossil record of a particular time was exposed or eroded, preserved in sediment or obliterated by metamorphism, buried far below the surface or submerged underwater.

Fossilization is a rare event in nature. But sometimes a flood or load of volcanic ash covers a fresh carcass with sediment. Then soft body parts usually decompose, leaving mostly bones and teeth. Mineral-laden water fills the bone pores, and eventually sediments around the skeleton become stratified rock. The shape of the skeleton, whether dinosaur jaw or delicate bird wing,

"The earth's crust is a sepulchre. Its sediments, which are ten miles thick, are full of the relics of plants and animals" (Emmons, *Report of Professor Emmons on His Geological Survey of North Carolina*, 1852).

Paleozoic Mazon Creek fossils from Illinois. Courtesy of Rosamond Purcell.

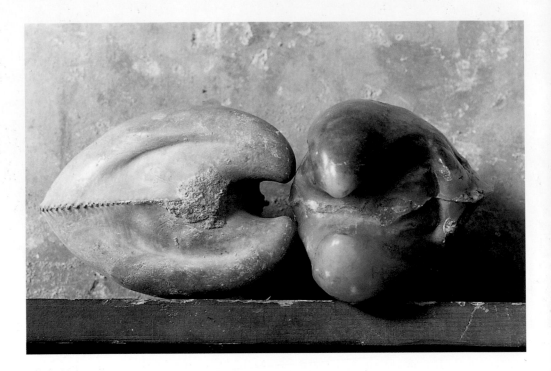

Cardium spillmani fossils. A fossil is a remnant of life preserved in stone. Shells of this Miocene clam were preserved as molds when the actual shell leached away, leaving only the shape of the internal body cavity. Courtesy of Rosamond Purcell.

is preserved in stone. Geologic forces may uplift and erode the rock, perhaps even move it to another hemisphere, until one day it is exposed to view.

Remains of organisms are preserved in other ways as well. Insects and spiders become trapped in a drop of sap, which turns to stony amber. Mammoths are preserved in ice and tar. Trace fossils give evidence of life: ghost shrimp burrows, salamander tracks, and dinosaur droppings announce that an animal passed by. Shells of animals like the Miocene clam *Cardium* are preserved as molds. Beachcombers who find the well-preserved molds of these ancient bivalves on North Carolina beaches sometimes mistake them for recent turtle beaks. Some shells may simply remain intact, unmineralized.

The meaning of fossils became apparent only in the last three centuries. Prior to the 1700s, "fossil" applied to anything found underground. John Lawson thought of fossils as "Earths, shells, stones, Mettalls, Minerals, stratas, paints, Phisicall Earths . . . subterranean matters." In 1743, naturalist Mark Catesby recorded fossil shells and marine vertebrates "imbedded a great Depth in the Earth," exposed in eroded riverbanks in Virginia. Catesby also reported that "at a place in Carolina [probably South Carolina] called Stono, was dug out of the Earth three or four teeth of a large Animal." Africans in the area identified the fossil teeth as "the Grinders of an Elephant"—an identification that, according to paleontologist George Gaylord Simpson, was probably the first technical identification of an American fossil vertebrate—the elephant being a southern mammoth.

William Bartram was one of the first to describe the rich fossil beds of the Cape Fear River near Elizabethtown where, in 1777, he collected fossil sea shells, wood "transmuted," and shark teeth, which the locals called "birds' bills." Later collectors would find dinosaur fossils here.

In the early 1800s, collectors discovered many fossils in eastern North Carolina. Growing collections in North America and Europe fueled debate about the origin of the species suggested by fossils. Before Charles Darwin forever altered the scientific worldview with the publication of *Origin of Species* in 1859, geologists here and abroad struggled to reconcile Scripture-based theory with the new data. Some of the beliefs and theories that prevailed before Darwin still inform the cultural dialog today.

By the 1820s, most geologists agreed that the wealth of newly described fossils proved that species became extinct and were replaced by new species. Extinct species of fish, mollusks, turtles, sea lizards, and crocodiles were thought by many to have perished in the great "Deluge" described in Genesis.

Denison Olmsted, professor of geology at the University of North Carolina, believed in catastrophism, the extinction of species by cataclysmic events. In a report to the state legislature in 1825, he noted that the excavation of the Beaufort Canal exposed a "thin layer of Sand, full of sea shells and the remains of land animals, particularly of the Mammoth, or fossil Elephant. Along with a profusion of shells, in perfect preservation, there are not unfrequently thrown out, huge teeth, vertebrae, and skeletons, more or less entire, of a gigantic race of animals, which no doubt, were buried here by that great catastrophe which also shut out the ocean far eastward of its original borders." Olmsted was probably referring to Noah's flood.

As the ages of rock layers became known, geologists realized that fossil

Glossus lunulata, mid-Miocene fossil snails from Belgium. A profusion of fossils were identified in the early nineteenth century in Europe and America. Naturalists endeavored to explain the origin and extinction of species in the fossil record. Courtesy of Rosamond Purcell.

animals existed in vastly separated time periods. Louis Agassiz, America's foremost naturalist of the time, believed that God repeatedly destroyed creation in a series of great cataclysms. Each geologic era signaled a fresh start, with a new creation of species.

Agassiz's belief was shared by Elisha Mitchell, who followed Olmsted in the geology chair at the university. Mitchell, an ordained minister, believed that the "last great catastrophe . . . was the deluge recorded by Moses." But he taught his students that the Bible was not about "physical truths" like geology, astronomy, and chemistry. He cautioned that a literal interpretation of Scripture could interfere with "the freedom of philosophical inquiry," a pursuit that Mitchell felt was not in conflict with religious truth.

Mitchell probed for physical evidence of each period of creation. To help determine the geologic eras represented in eastern North Carolina, he sent two boxes of fossil shells to Philadelphia for identification by famed conchologist Timothy A. Conrad. Mitchell lamented the lack of resources he had to invest in such important but consuming work. "The University," he explained, "is at a distance of between 70 and 80 miles in a straight line, from the nearest fossil shell." The prolific Triassic fossil beds nearby were yet to be discovered.

While Mitchell and his students labored in Chapel Hill, the world's foremost geologist, Englishman Charles Lyell, was exploring the rich Castle Hayne formation near Wilmington. Lyell stirred a revolution in the scientific world in the 1830s with his *Principles of Geology*. In it he described Earth's geologic processes as basically uniform throughout history—not catastrophic, one-time events, but forces in constant motion, like erosion, sedimentation, and volcanism. The first volume of his *Principles* was part of the shipboard library of young Charles Darwin, who later credited his good friend Lyell with opening his mind to new modes of thinking.

Elisha Mitchell, geology professor at the University of North Carolina, believed that catastrophes, including Noah's flood, were the cause of extinctions. Courtesy of the North Carolina Division of Archives and History.

Lyell supported his theories with solid fieldwork—including fossil collecting in eastern North Carolina along the Neuse River and Lewis Creek. In Philadelphia, he examined a fossil horse tooth from eastern North Carolina in the collection of T. A. Conrad, who also collected fossils in the state. Lyell compared the North Carolina fossil to one found by Darwin in South America, which was deposited near fossils of mastodon and giant ground sloth. Impressed with the similarities of the faunas of the two Americas, Lyell wrote, "It is a fact well worthy of attention that in the southern states of the Union so many extinct quadrupeds, such as the mastodon, elephant, megatherium, mylodon [both ground sloths], and horse" were also found in South America. Because these large animals were found with recent marine fossil shells "as Mr. Darwin has shown to be the case in the Pampas," Lyell thought it unlikely that their extinctions were caused by humans. "It is not the huge beasts alone," he reasoned, "but quadrupeds as small as the rat, which have become extinct in South America within the same period."

Before Darwin's 1859 publication, most geologists believed in essential-

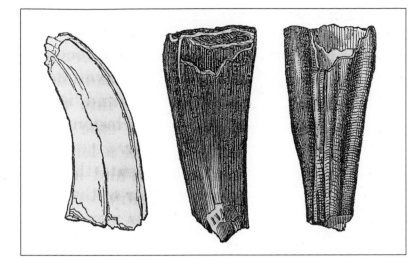

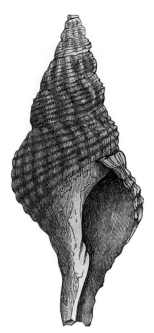

ism. Charles Lyell championed the idea, which held that species were constant over time, changing in certain characteristics but remaining essentially the way God created them. Extinct species were followed by replacement species of the same type, arising by special creation. Realizing that some species did not perish by catastrophe, Elisha Mitchell speculated that species may receive "the seeds of decay and dissolution" at the time of their creation, so they simply become extinct when their internal biological clock quits ticking. He was wrong, but well within the bounds of scientific thought at the time.

Also an essentialist, Ebenezer Emmons became North Carolina's first state geologist in 1851. A colleague of Lyell, Emmons often referred to the Englishman's theories in his reports to the North Carolina General Assembly. Emmons instructed the state legislators in the theory of French naturalist Georges Cuvier, which said that all animals, living and extinct, can be classified as one of four types—Vertebrata, Mollusca, Articulata, and Radiata. The types were defined by the "plan of creation" that continued unbroken to the present day. "All extinct animals are constructed upon one of the four leading types which now prevail. . . . It is the plan then, which really tells all this, or makes it possible to compare and infer with certainty."

Like Mitchell, Emmons was frustrated by poor resources, particularly lack of access to the work of other naturalists. He feared that he was repeating work that had already been done: "It may turn out that many species which have been marked as new, will prove to be old ones already described." Emmons believed he had found a new genus of fossils in Montgomery County rocks that were older than any fossil-bearing rocks known at the time. He believed the fossils to be two species of corals, which he named *Paleotrochis major* and *Paleotrochis minor*. He thought this find would please the people who claimed "with an obvious feeling of pride, that North-Carolina has the highest peak east of the Rocky mountain range. It will no doubt be amusing

Fossil horse tooth (*above left*), Bladen County. English geologist Charles Lyell used fossils to determine the ages of rock layers. He discovered Eocene fossils in eastern North Carolina, and compared a fossil horse tooth from the state with one found by his friend Charles Darwin in South America. From Emmons, *Report of the North Carolina Geological Survey*, 1858.

Miocene mollusk (*above right*), *Fasciolaria sparrowi*, Beaufort County. Many scientists in the early nineteenth century believed that all animals, living and extinct, could be classified as one of four types—Vertebrata, Mollusca, Articulata, and Radiata. From in Emmons, *Report of the North Carolina Geological Survey*, 1858.

Paleotrochis major (*right*), found in Montgomery County by Ebenezer Emmons, was thought to be the oldest fossil known. Geologists later determined that the "fossil" was a naturally occurring property of the rock. Courtesy of Rosamond Purcell.

Washington Caruthers Kerr found the state's first documented dinosaur fossils (*below*), named *Hypsibema crassicanda*, near the Cape Fear River in 1869. These *Hypsibema* vertebrae were found in 1998 near Kerr's dig sites. Courtesy of Jim Page.

Paleotrochis major

Paleotrochis minor

to others, should I claim for North-Carolina, the honor of being the birth place of the oldest inhabitants of this globe." Unfortunately for state pride, Emmons's "fossils" later were determined to be naturally occurring rock formations, called spherulites, embedded in rhyolitic volcanic glass.

Despite some errors, Emmons's 100-page documentation of North Carolina fossils contributed to the growing scientific literature on North American fossils. Emmons's discoveries enlarged the state's early geological collection with fossils of crocodiles, mosasaurs (flippered marine lizards), and turtles of the Cretaceous period, along with mastodons, horses, whales, sharks and bony fishes, mollusks, and echinoderms.

A careful fieldworker, Emmons cautioned North Carolina fossil collectors to stick with objective science, but to use the plan of creation as a theoretical framework. "Observation is the way, but the plan of creation makes it possible to deduce a connected history of the past from the dead races." Like Mitchell, Emmons looked to nature for scientific truth. He seemed ready to revise theory as new data came to light, writing in 1856, "There is really no conflict of old with new facts; the conflict is with the new facts and old opinions. . . . Geologists erred in limiting nature. They introduced into science a dogma, which she repudiates."

As fossil evidence mounted, essentialists found it harder to reconcile their worldview with objective science. A year before Darwin's Origin of Species was published, Emmons expressed his generation's frustration over the puzzle of extinction and survival. "We are left to speculate on probabilities, without being able to arrive at satisfactory conclusions."

Emmons died in 1863, before the controversy surrounding Origin of Species hit home in war-torn North Carolina. But the writings of Emmons's successor in the post of North Carolina state geologist, Washington Caruthers Kerr, showed the influence of a decade of debate surrounding Darwin's theory of evolution by natural selection. After initial resistance in the scientific community, most naturalists accepted the idea of the unguided change of species by the 1870s, though some American scientists found it hard to abandon the requirement of a divine plan.

Kerr (who had studied under Agassiz at Harvard) accepted the Darwinian idea that change in species was directional—not cyclical or based on a constant type, as his professors had taught. But Kerr stopped short of a strictly secular interpretation of Darwinism. Kerr believed that an established plan, aimed at progress, guided the development of new species. "Old forms of life will have passed away and new ones occupied their places. Life forms, whose successive developments have left their impress upon the rocks, have . . . undergone great changes from age to age. These changes have been in certain determinate directions, involve more than mere change, and indicate also progress," Kerr wrote in his 1875 report to the state legislature.

Kerr collected fossils with a passion. He revisited the Eocene formation near Wilmington that Lyell first described in 1842, and published a review of

North Carolina's fossil shells by T. A. Conrad, including descriptions of new species found by Kerr. In 1869, Kerr found the state's first dinosaur fossils at Emmons's dig site in Sampson County. He sent ten bones for identification to famed dinosaur paleontologist Edward D. Cope at the Philadelphia Academy of Sciences, who named the animal *Hypsibema crassicauda*, "thick-tailed high-strider."

Kerr collected fossil whale bones exposed by creek erosion in several places in the Coastal Plain. One of his finds, *Mesoteras kerrianus*, was named by Cope for "Prof. W. C. Kerr, who has vitalized the State Survey and is prosecuting it with great advantage to all the branches of science that lies within its scope."

Cope himself was a controversial naturalist who spent time in North Carolina studying fishes. One of a dwindling number of biologists who resisted Darwin's theory, Cope led a brief revival of interest in Lamarck's early-1800s theory of evolution by acquired characteristics. However, Darwin's theory of evolution by natural selection eventually won consensus among biologists around the globe, even dyed-in-the-wool essentialists like Charles Lyell. With modification, it remains a basic tenet of the natural sciences today. Evolution is the organizing principle of systematics, the basis for classification of the zoological collections of the North Carolina State Museum of Natural Sciences.

Shark teeth and whale fossils were displayed at the State Museum by the 1890s. Museum curator H. H. Brimley kept a portrait of Darwin in his workroom and actively collected fossils. In 1909, Major H. T. Patterson of the U.S. Army Corps of Engineers contacted the State Museum about an interesting group of bones unearthed by dredges digging the Intracoastal Waterway. Brimley sent Thomas Addickes to excavate in the area. Addickes reported, "We secured most of the skull, all the teeth, both tusks and the principal

Mammoth sketch, by H. H. Brimley, State Museum curator (*left*); mastodon bones in State Museum workroom (*right*). Museum reconstructions of extinct vertebrates allowed artists to imagine how the flesh-and-blood animal appeared. North Carolina State Museum Archives.

bones of the right hind leg and the left fore leg, of a medium-sized mastodon." The 15,000-year-old mastodon skeleton, a favorite museum exhibit for fifty years, moved visitors to wonder at its immense size, two-foot tusks, and the irrefutable fact that it once roamed the same North Carolina landscape where they farmed and fished.

H. H. Brimley contemplated the process of the mastodon's fossilization: "In the swamps and morasses of the tide-water section of North Carolina, every opportunity was afforded for an animal to lose its life by bogging down in the peaty soil . . . perhaps in mud so soft and deep that the whole animal would, in a very short time, become completely engulfed and sink below the normal water-level, so that the skeleton would remain intact for an almost unlimited period of time—for the fortunate museum collector who happened to discover it."

However, museum displays were quiet on the subject of evolution, sensitive to the religious beliefs of some state citizens. There is no record of a direct interpretation of evolution in the State Museum prior to the twenty-first century, although museum associate C. S. Brimley lectured on "the descent of the various groups of animals" at the Biltmore Forest School at the turn of the twentieth century. Evolutionary theory is described in the state's public school biology textbooks, though its inclusion was challenged as recently as 1997 in the North Carolina General Assembly.

The 110-million-year-old *Acrocanthosaurus atokensis* dinosaur skeleton on display in the State Museum bears evidence of the tough life of a predatory dinosaur. The skeleton retains the scars of an infected shoulder, broken bones, and bite wounds on its skull. Courtesy of Peter Damroth.

The State Museum's paleontology program expanded in the 1990s, adding staff paleontologists who revisited earlier dig sites in the state. Their findings, along with the fossil record accumulated since Olmsted's day, are interpreted in five habitat dioramas that model North Carolina's prehistoric past. Details about these early environments come from many sources. Wind-dispersed pollen and seeds in soil cores from peat bogs indicate the presence of open forests. Fossilized leaves and wood suggest average temperatures and rainfall. Evidence of nesting sites shows how some animals cared for their young.

The size, weight, and configuration of bones suggest the type and amount of food required by an animal, and fossilized droppings may preserve components of an animal's diet. How fast an animal moved can sometimes be determined with a good set of preserved tracks. Marks on animal bones bear witness to conflicts between predator and prey. The *Acrocanthosaurus* skeleton in the State Museum's collection carries the marks of the violent life of a predatory dinosaur: a severely infected shoulder, several broken and re-healed ribs, a broken toe, and bite wounds on its skull.

Paleontologists fill in gaps in North Carolina's fossil record with what is known from adjoining areas. As continental plates realigned, southeastern North America was linked at various times by land bridges or close contact with early Africa, Europe, and South America. Using the fossil record and clues from Earth's geologic history, paleontologists pieced together these images of life in North Carolina throughout time.

Precambrian Era—4.6 Billion to 543 Million Years Ago

Before life began, the molten surface of Earth was battered by a gradually slackening rain of giant comets and asteroids. The young sun penetrated an oxygenless atmosphere with only 70 percent of the light of later eras. For over a billion years, Earth's surface cooled as rainwater filled the oceans.

Within that first billion years, organic molecules began to form from simple chemicals. Somehow—perhaps through chemical self-assembly—those molecules began replicating themselves. Simple cells emerged. Primordial bacteria and the strange archaea flourished around sulfurous hot springs. Scientists study their descendants, which survive in exotic habitats like the steam vents in Yellowstone National Park.

Around 3.5 billion years ago, great mats of blue-green algae spread out across the seas. Called stromatolites, these masses of cells were able to make food out of inorganic chemicals and sunlight. Mutant offspring of these earliest photosynthetic bacteria began to produce oxygen, a new gas in the atmosphere that poisoned the air for older life forms. Stromatolites still exist in a few marine environments.

Another billion and a half years passed, and the oxygen in Earth's atmosphere increased. Around two billion years ago, complex cells arose, equipped

with nuclei and specialized parts capable of exploiting the new oxygen resource. Nuclei enabled cells to be genetically different from each other, and diversity mushroomed. Sexual reproduction arose around 1.2 billion years ago, just as the oldest rocks in North Carolina were being formed. Around 800 million years ago, multicelled organisms began to diversify into scores of new species: sea pens, sponges, worms, and others. Earth was nearly four billion years old when multicellular life began to flourish.

North Carolina fossils of the soft-bodied organisms of the Precambrian era are extremely rare. Mountain-building episodes altered early rocks, leaving only a few fossils, including the sea pen, *Pteridinium*. Because most Precambrian rocks were recycled into Earth's crust by tectonic activity, little evidence remains of early life in North Carolina.

Paleozoic Era—543 to 245 Million Years Ago

Life in Paleozoic seas diversified during a burst of evolution. Although the igneous and metamorphic rocks from this era in North Carolina offer no hint of life, the global fossil record documents a multitude of new aquatic animals, including most major groups of invertebrates. The trilobite, a ubiquitous early arthropod, populated the seas for 300 million years. Fishes, the first animals with backbones, arose later in the era. Jawless fish developed a backbone and a skull—equipment that gave all the vertebrates an unparalleled freedom of movement. Predatory fish evolved in many shapes and sizes—fish with jaws, armored fish, and sharks.

Plants began to colonize the land about 400 million years ago. They spread quickly, outfitted with innovative structures—roots, stems, a waterproof "skin," and internal transport tubes. Ferns, seed-bearing plants, and 100-foot-tall scale trees created lush swamps and dense forests. Amphibians were the first vertebrates to truly inhabit the land, though most returned to water to lay their jelly-like eggs. Reptiles could both live and breed on land, thanks to the innovation of eggs with shells.

Some Paleozoic animals grew to tremendous size, including giant armored fish, two-foot-long dragonflies, and dinner-plate-size spiders. Their way of life ended about 245 million years ago in the most devastating mass extinction of all time. More than 95 percent of Earth's species vanished, opening the door to a new era.

Triassic Period (Mesozoic Era) — 245 to 213 Million Years Ago

Some of the Paleozoic survivors gave rise to new groups — turtles, mammals, dinosaurs, and birds. Reptiles dominated the long Mesozoic era: huge marine reptiles preyed on fish, pterosaurs dove for fish from above, and dinosaurs ruled the land. In the early Mesozoic, eastern North America and Africa were joined as the supercontinent Pangaea. Warm, humid North Carolina was close to the equator, about a mile above sea level. Large valleys began to open as the continental plates pulled apart. Tall conifers spread across these rift valleys, skirting large lakes rimmed with horsetails, cycads, and ferns.

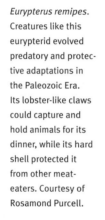

Eurypterus remipes. Creatures like this eurypterid evolved predatory and protective adaptations in the Paleozoic Era. Its lobster-like claws could capture and hold animals for its dinner, while its hard shell protected it from other meat-eaters. Courtesy of Rosamond Purcell.

The rift basins of Piedmont North Carolina preserved some of the best Triassic fossils on the East Coast. Fern fossils are abundant, indicating that ferns were a dominant part of a humid and marshy landscape. Fossils of plants, insects, amphibians, and phytosaurs abound in rift lake deposits in the basins of the Dan River and Deep River. The Dan River site preserved fossils of the oldest known true water bugs and some of the oldest known true flies.

Waterholes held danger for large plant-eaters, which were easy prey for fierce phytosaurs and crocodile-like rauisuchids. These carnivores were primitive members of a reptile group called archosaurs, or "ruling reptiles,"

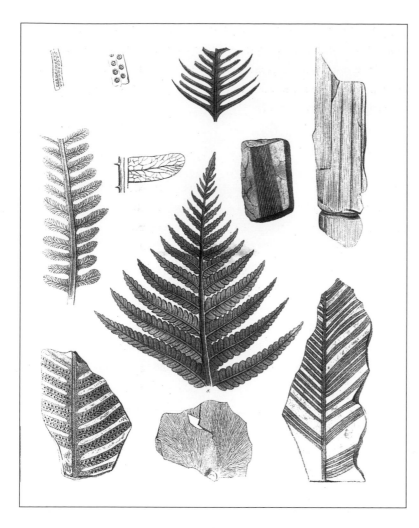

Triassic ferns, 213–245 million years old. Well-preserved Triassic fossils are found in North Carolina's fine-grained shales. These ferns from the Deep River at Jones' Falls were collected and described by Professor Emmons in the 1850s. From Emmons, *Geological Report of the Midland Counties of North Carolina*, 1856.

which reigned for 20 million years. Another group of archosaurs, the dinosaurs, appeared by the end of the Triassic period. The oldest fossils of dinosaurs in North Carolina are the 220-million-year-old teeth of a plant-eater, *Pekinosaurus olsoni*, found in Montgomery County.

Ecosystems faltered at the end of the Triassic, and extinctions took out the rauisuchids, phytosaurs, and other large animals. Some archosaurs survived, including crocodiles, pterosaurs, and dinosaurs.

Jurassic Period (Mesozoic Era) — 213 to 144 Million Years Ago

Eventually Africa pulled away from North America, leaving Piedmont rocks touching the seashore. The high Appalachians eroded swiftly, piling sediments on the coast. Today, the 70-million-year fossil record of Jurassic North Carolina is buried under thousands of feet of younger sediments off Cape Hatteras.

From the fossil record in other regions, paleontologists surmise that the period was warm, allowing plants to spread to the polar regions. Earth's

Jurassic atmosphere contained much more carbon dioxide than does its current atmosphere. Dinosaurs diversified and giant leaf-eating sauropod dinosaurs became abundant. Mammals remained small, but diversified more rapidly than dinosaurs. Birds evolved, joining pterosaurs in the skies. Ichthyosaurs (reptiles that looked like dolphins) and plesiosaurs populated the seas.

Cretaceous Period (Mesozoic Era)—144 to 65 Million Years Ago

The supercontinent Pangaea finally split into the continents we know today, although some of them, like Australia and Antarctica, remained connected. Land plants and animals became isolated, and more diverse, as the continents separated. The climate cooled, especially at the poles. North American dinosaur species tended to be smaller and more specialized than earlier species.

A shallow inland sea ran between the Rockies and the Appalachians, isolating species on both sides of North America. One of the largest deltas in eastern North America spanned the coast of the Carolinas. Most Cretaceous fossils found in North Carolina, including crocodiles, turtles, fishes, and dinosaurs, come from the sediments of the Cape Fear River basin in this area, where W. C. Kerr found specimens of dinosaur bones in 1869. Originally described as the large plant-eater, *Hypsibema*, the bones are now known to be those of several different dinosaurs. Four types of dinosaurs have been found in the area. Digging in the same locale in 1998, Vince Schneider, State Museum curator in paleontology, led a team that discovered dozens of fossil crocodiles, dinosaurs, turtles, sharks, and mosasaurs, many of which are now on exhibit at the State Museum.

A dinosaur arm bone found in the area belonged to one of the largest duck-billed dinosaurs ever found in North America. Duck-billed dinosaurs gathered by the thousands in enormous, smelly waterside nesting grounds. Adults incubated the eggs for several weeks, aided by rotting vegetation, and

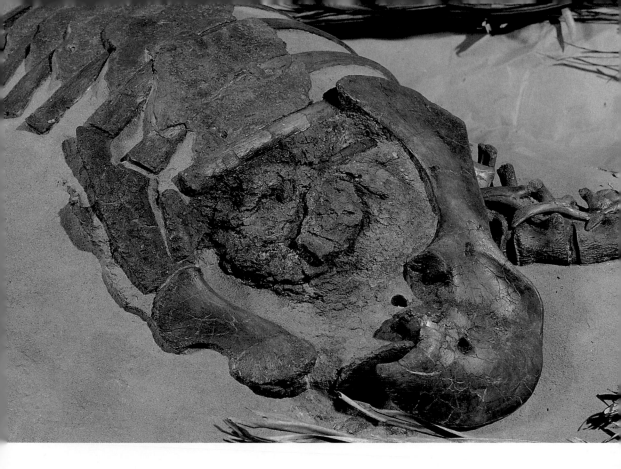

fed the growing hatchlings regurgitated meals. Hatchlings were easy prey for flesh-eating dinosaurs and crocodiles.

North Carolina fossils of the dinosaur *Acrocanthosaurus* probably lie beneath a mile of younger sediments at the bottom of Albemarle and Pamlico Sounds. But *Acrocanthosaurus* fossils from Maryland, Texas, and Oklahoma indicate that this animal was once the most common large carnivorous dinosaur in the region—the "Terror of the South"—one who preyed on dinosaurs much larger than itself. Adult dinosaurs were also prey for animals like *Deinosuchus*, a crocodile that grew to 40 feet and weighed several tons.

As with all animals, the movements of Cretaceous dinosaurs—how they hunted, how they fled—were dictated by body metabolism. A fossil in the State Museum's collection, a 66-million-year-old thescelosaur from western North America, may hold the clue to the metabolic rate of some dinosaurs. Using medical technology, State Museum curator in paleontology Dale Russell, and colleagues at North Carolina State University and in Oregon discovered the impression of a heart in the well-preserved skeleton of the plant-eating dinosaur nicknamed "Willo." Initial examination of the dinosaur's heart and associated tissues indicated that its metabolism was more avian than reptilian. If so, Willo and others of its kind would be capable of greater endurance in sustained activity—more like a bird or mammal than a lizard or turtle. As a result, Willo might have been able to outrun or outdistance a threatening predator.

Thescelosaurus neglectus ("Willo"), 66 million years old, South Dakota. The most complete thescelosaur known, this specimen has the best-preserved skull of a small plant-eater ever found in North America. Preserved impressions of tissue in the chest area (possibly a heart) may indicate that it had a metabolic rate closer to birds and mammals than to reptiles. Courtesy of Jim Page.

Plant-eaters fed on the flowering plants that arose in the late Cretaceous, which spread quickly in areas defoliated by browsing dinosaurs. Dinosaur droppings helped disperse flower seeds in fertile territory. Sycamores, magnolias, and palmettos began to dominate ecosystems where ferns, cycads, and conifers once stood. According to Russell, natural areas in eastern North Carolina today resemble in some ways the Cretaceous world that existed on the shores of North America's inland sea. Close relatives of Cretaceous plants and animals inhabit North Carolina's swamps, coastal marshes, and estuaries. Longnose gars, soft-shelled turtles, alligators, and opossums live among bald cypresses, magnolias, palmettos, and water ferns much as their ancestors lived in the day of the dinosaur. Coupled with the fossil record, this living environment sheds light on what life was like in Cretaceous North Carolina.

The period ended 65 million years ago when a comet slammed into the area of Mexico's Yucatan Peninsula. Global environmental catastrophe ensued—tidal waves, intense heat, and darkness were followed by a cooler period of acid rain. Green plants failed on land and sea, resulting in mass starvation in ecosystems around the world. All land animals larger than a dog disappeared, and many aquatic species became extinct. A few scavengers survived, and some plants were able to live on food stored in seeds or roots.

Tertiary Period (Cenozoic Era)—65 to 1.8 Million Years Ago

Ten million years after the comet collided with Earth, life surged again, and birds and mammals became prominent life forms. Tropical forests covered Earth; primates, bats, carnivores, and cetaceans flourished in the absence of dinosaurs. Wolf-like, fish-eating mammals gave rise to whales, dolphins, and porpoises. The ancient whale *Ambulocetus natans* lived in the ocean and bred on land. It could metabolize seawater, a major step in whale evolution. The Coastal Plain preserves an abundance of fossilized whales from the Tertiary period.

A shallow, warm sea covered eastern North Carolina in the Eocene epoch 55 to 35 million years ago. At times, the eastern continental shoreline reached almost to present-day Raleigh. Colonial animals called bryozoans formed large underwater reefs. When the climate cooled, rain forests yielded to grasslands, and hoofed mammals appeared. Bony fish diversified.

By the Pliocene epoch 10 million years ago, most of the bony fish we know today had appeared. Fossils of bluefish, bonito, pufferfish, goosefish, sea bass, and sea robin are common in the rich Pliocene fossil record of eastern North Carolina. Sharks, whales, walruses, and seals fed on the abundant fish populations in the cool, deep waters off the coast. Teeth of a 40-foot *Carcha-*

Museum diorama of Cretaceous North Carolina. On the banks of the Cape Fear River, a meat-eating *Albertosaurus* crashes through underbrush, taking an *Edmontosaurus* duckbill dinosaur by surprise. An alarm spreads through the duckbill nesting colony. Animals scatter in all directions. Courtesy of Peter Damroth.

rodon megalodon, ancestor to the great white shark, were described by Emmons as the most common of fossil teeth in the Cape Fear area. From the size of the 6-inch-long teeth, Emmons deduced, "It must have been one of the largest and most formidable animals of the ocean . . . the most terrific and irresistible of the predaceous monsters of the ancient deep." Among the great diversity of fossil fishes found at the Lee Creek phosphate mine in coastal Aurora, North Carolina, was a set of 25 teeth from one *C. megalodon* shark. Becky Hines, who discovered the fossils in 1992, offered them to the State Museum, where the "monster's" teeth are displayed in a six-foot-wide reconstructed jaw.

Quaternary Period (Cenozoic Era) — 1.8 Million Years Ago to Today

In northern North America, glaciers advanced and retreated along with fluctuations in the global climate during the Pleistocene epoch. Habitats in North Carolina changed from woodlands during glacial periods to open savannas in warmer times. When sea level fell in glacial periods, the state's coastline extended eastward, exposing most of the continental shelf.

Land bridges emerged, allowing animals to move between North America, South America, and Asia. North American horses, mastodons, rodents, cats, and dogs went south. The North American tapir still survives in South America. Ground sloths, armadillos, anteaters, and glyptodonts came north.

Remains of giant ground sloths, giant beavers, mastodons, and mammoths of the Pleistocene were preserved in Coastal Plain sediments. Since the early nineteenth century, collectors have found ribs, jaws, and teeth of mastodons in eastern North Carolina sites. Workers at a construction site in Wilmington inadvertently unearthed the burial ground of several giant ground sloths in 1991. Fossil collector Mike Young was looking for shark teeth on the site when he discovered the sloth bones, which were deposited on the beach or in the Cape Fear River estuary about 1.5 million years ago. Schneider and a team of volunteers from the State Museum and the Cape Fear Museum excavated a thousand bones at the site, the remains of a small group of one of the oldest ground sloth species in North America. The rearticulated skeleton of one of the sloths is nearly 22 feet long. Powerful arms and long claws allowed the giant ground sloth to defend itself from the North American saber-toothed cat.

As glaciers retreated at the end of the Pleistocene 12,000 years ago, sea level rose and North Carolina's barrier islands began to form. Humans entered the continent on a land bridge connecting Asia and Alaska. The change in climate coupled with the arrival of predatory humans were possible causes for the extinction of three-fourths of the large land mammals in North America.

The past two hundred years of fossil collecting have revealed a history of life more diverse and complex than early paleontologists could have imagined. Fossils allow a glimpse of another world with its own atmosphere,

(opposite)
Teeth of *Carcharodon megalodon*, 10 million years old, Beaufort County. The ancestor of the great white shark, *Carcharodon megalodon* hunted in Pliocene seas. Its fossil teeth are fairly common in the Coastal Plain. A set of 25 teeth from one animal is on display in the State Museum. Courtesy of Rosamond Purcell.

Museum diorama of Pliocene North Carolina. A seal feasts on a school of bonito in cool water off the North Carolina coast. Large schools of bluefish and anchovy swim away from the disturbance made by the seal. Courtesy of Chase Studio.

ecosystems, and physical stresses. Knowledge of how life survives in times of stress—mass extinction, restricted resources, altered climate—may suggest what could happen in our own world when similar events occur. Fossil evidence also gives biologists the data they need to construct family trees for living species—the evolutionary relationships that are the foundation of systematics collections.

FOUR Invertebrates

Invertebrates are animals without backbones. Just about every multicelled creature falls into this category—about 97% of all animal species on Earth are invertebrates. Newly discovered species are added to the more than a million and a half named species as fast as taxonomists can describe them. An estimated 10 to 30 million invertebrate species have yet to be discovered.

Ordering this great diversity of species has always been a difficult challenge. The great taxonomist Linnaeus, so discerning in the classification of plants and vertebrates, lumped most invertebrates together as "Worms." Today, taxonomists use up to 30 phyla to organize the realm of invertebrates. Millipedes (phylum Arthropoda) and leeches (phylum Annelida) have segmented, elongated bodies. River mussels (phylum Mollusca) possess an organ that secretes a hard shell. Spiders and horseshoe crabs (two classes in phylum Arthropoda) have jointed legs and jawlike chelicerae; mayfly larvae and dragonfly nymphs (two orders in class Insecta, phylum Arthropoda) metamorphose into six-legged adults with three main body sections.

Invertebrates are everywhere, from the spruce-fir moss spider atop Mount Mitchell to the Carolina octopus swimming off the continental slope. Colorful millipedes, lampshade spiders, and mayflies abound in the mountains. Crayfishes, river mussels, dragonfly nymphs, and leeches play essential ecological roles in all of the state's waters.

Early naturalists were daunted by this enormous assemblage of animals. John Lawson in 1709 described the state's insects as "too numerous to relate

"One Cockle in Carolina is as big as five or six in England. These make an excellent strong Broth, and eat well" (Lawson, *A New Voyage to Carolina*, 1709).

Northern quahogs, *Mercenaria mercenaria*. Courtesy of Rosamond Purcell.

A lande Crab.

here, this Country affording innumerable Quantities thereof; as the Flying-Stags with Horns, Beetles, Butterflies, Grashoppers, Locust, and several hundreds of uncouth Shapes." Lawson found shellfish—crayfish, crabs, and mollusks—worth describing because of their food value. The colonists esteemed the "Man of Noses" mollusk for "increasing Vigour in Men, and making barren Women fruitful."

The study of mollusks was prized by nineteenth-century geologists; fossil mollusks helped date rock strata around the world. University of North Carolina Professor Elisha Mitchell instructed his students in 1842 that a geologist should be "well versed in the science of conchology." Fossil mollusks from the Piedmont and Coastal Plain of North Carolina were described and displayed in the state's early Cabinet of Minerals.

However, most invertebrate research has focused on economically important animals—edible species and agricultural pests. A collection in "economic entomology" was the first invertebrate display in the State Museum, along with a few marine mollusk shells. Today the museum's systematics collections concentrate on ecologically significant groups with poorly developed classifications. Millipedes, crayfish, and mollusks all figure largely in the food web of many ecosystems, and curators are discovering that these groups are more diverse than once thought.

Phylum Arthropoda: Millipedes

Many-legged millipedes are the unsung heroes of the forest ecosystem, recycling decaying plants in Herculean amounts. A millipede ingests dead plant material, extracts the nutrients, and excretes the remainder. This action repeated by dozens of species and hundreds of individuals in one acre of woods adds up to a major recycling effort. In tropical areas with few earthworms, millipedes may be the major recyclers.

The success of millipedes as a group is undisputed; the oldest living group of land arthropods, they have survived for 420 million years. Taxonomists have classified 8,000 or so millipede species in 15 orders, and suspect that only about one-tenth of the world's millipedes are known to science. In a 30-year study of millipede diversity, Rowland Shelley, State Museum curator of terrestrial invertebrates, discovered 160 new species of millipedes and reclassified major groups of this little-known class of animals.

Sigmoria sp., millipede from the Mountain region of North Carolina. This kind of millipede displays color markings that warn predators of the toxicity of their defensive secretions. Courtesy of Harry Ellis.

Shelley has shed light on the evolutionary success of millipedes by looking at mimicry in the Mountain region of North Carolina. Ancient North Carolina mountain ranges were a millipede stronghold, home to a group of animals that would later diversify into a dizzying number of species that now occur to Mexico and the Pacific Coast. Today the southern Appalachians

claim one of the most diverse millipede faunas in the world. Decked out in bright red, blue, orange, white, and yellow markings, many mountain millipedes are easy for predators to find, but they usually make a terrible meal. Many of the more than 100 mountain millipede species in North Carolina and surrounding states produce toxic hydrogen cyanide along with benzaldehyde, a fragrant almond-scented compound.

Millipede markings gave Shelley an insight into mimicry, an evolutionary adaptation. Comparing countless preserved specimens and hunting forest millipedes in back-road mountain coves, he found that many species in a given location look alike. Each species bears a sort of neighborhood coat of arms—a distinctive pattern of spots and stripes. In fact, each species may look more like its unrelated neighbors than its own kin in the next valley, which in turn has its own neighborhood style.

Each prey community, isolated by mountain terrain, seems to have evolved a local warning sign that predators learn to recognize. Predators may learn the pattern sooner if every foul-tasting millipede they encounter bears the same markings. The more animals that sport the message, the bigger the advertising impact. In terms of survival, a quicker learning curve for the predator means fewer millipedes sacrificed and less wasted energy for predators, who will retrain their sights on more palatable creatures. Both predator and prey benefit from this system.

Phylum Annelida: Leeches

Accurate identification of species helps biologists know who's who in the natural world. New species may be discovered in the field or in collections. Two unidentified leeches sat on the shelves of the State Museum invertebrate collection for years. In 1971, Shelley took them to leech specialist Roy Sawyer, then working in South Carolina. Together they determined that the nameless creatures were in fact members of a new terrestrial species, only the second recognized from North America. Inching along under rotting logs or wet leaf litter in Piedmont woods, *Haemopis septagon* had eluded scientists for years, although at 6.5 inches, it is the second largest leech in North America. Not a bloodsucker, *H. septagon* swallows earthworms, whole.

The state's aquatic leeches feed mainly on other invertebrates, including mosquito larvae and other leeches. They live in shallow, slow-moving waters of ponds and streams. A few marine species live in North Carolina's estuaries and sounds, sucking blood from fish.

The blood-sucking behavior of some leech species earned the animals a role in the state's folklore. Some Cherokee still tell of the great leech of Tlanusiyi, the spot where the Valley and Hiwassee Rivers converge at the mountain town of Murphy. As big as a house, the leech swept travelers into a deep hole in the river and left their remains with ears and noses missing.

Folk and professional medical practitioners relied on bloodletting leeches as therapy for a host of illnesses in the nineteenth century. As blood suckers, North Carolina native leeches were spurned for the more bloodthirsty European species, which could ingest an ounce of blood in one sitting. Anna Pritchard of Warrenton, North Carolina, documented local leech therapy in her diary entry of 1854: "Dr. H. said John must be leeched. Dr. P. sent Thompson to Dr. Fields & told him, if he could get no leeches there, to go to town and get some from Dr. H.—The boy came back about an hour, by sun, with a bottle of common leeches. (We had sent to the ponds that day but could get none.) We moistened John's breast with chicken blood, but they would not bite. Dr. H. said . . . he never used common leeches. (We sent the boy straight to town & and he got back a few minutes past 10 o'clock at night with a quinine bottle of German leeches. How glad I was to see them!)"

Phylum Arthropoda: Crayfishes

To study one of the largest freshwater invertebrates in North Carolina, a naturalist need only visit the nearest stream and turn over a rock. Chances are the rock will be sheltering a member of one of the 35 to 40 crayfish species native to the state. Once its hiding place is invaded, the animal will attempt an escape by darting backwards with astonishing speed and agility. A careless collector will meet its second line of defense—a pair of powerful claws with sharp-tipped fingers.

John Lawson found the baited-stick method Native Americans used for catching crayfish so effective that a short trip to a creek produced bushels of crayfish "which are as good, as any I ever eat." A nineteenth-century record for the red burrowing crayfish, *Cambarus carolinus*, noted that it was collected from "'Among the Cherokees', Indian Territory" in North Carolina, and was "called *Tsisgágili* by the Cherokee Indians." In Cherokee oral tradition, when the world was made, *Tsisgágili* was scorched bright red by the fiery young sun. His meat was spoiled, and the Cherokee would not eat it. But they believed the crayfish could bestow its strong grip on a young child if its pincers scratched the child's hand.

C. S. Brimley collected the red crayfish in the 1930s in Swain County's Little Tennessee River. His specimens were among the first noninsect invertebrates in the State Museum. Decades later, Horton H. Hobbs Jr., the legendary crayfish specialist known as "Old Man Crawfish," sought to confirm Brimley's record, but the collecting site had been flooded by Fontana Dam. Hobbs was the unassuming dean of American crayfish studies, and named about half of the species known in the Americas, including a third of those known in North Carolina. Hobbs's colleague, John E. Cooper, current curator of the State Museum's crustacean collection, added 12 native species and two introduced species to the state list.

More species of crayfishes occur in the southeastern United States than anywhere else. Seven described and several undescribed species are found

exclusively in North Carolina. One of them, *Cambarus catagius*, was known only from burrows in a lawn in Greensboro until recently. Cooper credits the state's diversity of crayfishes to North Carolina's geologic and hydrologic history. Populations isolated by changes in the environment account for some species found nowhere else.

Cooper discovered *Orconectes (Procericambarus) carolinensis*, a crayfish limited to distribution in the Neuse and Tar-Pamlico river basins of North Carolina. The only member of its subgenus known in the eastern Coastal Plain, it probably evolved from ancestral stock that once inhabited western river basins. The ancestral crayfish may have entered eastern drainages as streams changed course. Isolated from its progenitors in the west, *O. carolinensis* emerged as a distinct species.

Another crayfish, *Procambarus (Ortmannicus) braswelli*, is found only in the Waccamaw and Lumber river basins in the southeastern Coastal Plain. A primitive member of a group more common farther south, the Waccamaw crayfish is only distantly related to its geographically nearest crayfish relative. Lake Waccamaw, the largest of the ancient elliptical depressions called Carolina bays, is also home to several mussel and fish species found nowhere else. The lake may owe its endemic fauna to the unique chemistry of its water. Steep bluffs along the north shore of the lake are evidence of underlying calcareous deposits, which contribute to the lake's exceptionally high pH and alkalinity.

Cambarus dubius is one of seven true burrowing crayfishes found in the state. Sometimes tunneling far from surface water, some burrowing crayfishes may dig as much as 12 feet below ground before they reach groundwater. Their complex burrow systems serve them for life. Chimneys as high as one foot may mark the entrances to crayfish tunnels. Though biologists aren't sure why some crayfishes erect chimneys, the structures may serve as ventilator shafts, stirring air currents in the underground chambers. During

cold weather or drought, burrowers often cap their chimneys for climate control. Underground crayfish burrows provide shelter for other subterranean residents, from worms and pillbugs to fishes and amphibians.

Crayfishes supply the larder for many inhabitants of North Carolina's waters. They are eaten by more than 125 vertebrate species, including humans. Raccoons, otters, herons, ducks, and even hawks and owls eat their share. Turtles, alligators, some snakes, and some amphibians compete for them with such voracious crayfish eaters as bass, pickerel, and other fishes. Crayfishes, in turn, eat almost any kind of plant or animal matter they can find or catch. As a strand in the food webs of many creatures, crayfishes are essential to keeping the flow of energy and nutrients moving through freshwater ecosystems.

Phylum Mollusca: Freshwater Mussels

Once abundant in North Carolina, heaps of freshwater mussels have been found at riverside archeological sites. Native Americans harvested these bivalved shellfish for food, probably as a tasty supplement to their lean spring diet. At the thousand-year-old Donnaha site on the big bend of the Yadkin River near Winston-Salem, Siouan people left pits piled high with mussel shells. Tuscaroran people dined on freshwater mussels of the Roanoke River. Along the Yadkin and Eno Rivers, villagers left combs made from the shells of mussels; these tools were handy for smoothing the insides of coil-built pots and for scaling fish.

Later, Europeans came to the river banks with collecting bags, eager to add new species of mussels to the scientific literature. They discovered streams teeming with an unexpected diversity of mussels: one American river held five times the number of mussel species known in all of Europe. Biologists later determined that a third of the freshwater mussel species on Earth reside in eastern North America.

This cornucopia of undescribed animals set off a naming frenzy among nineteenth-century collectors on both sides of the Atlantic. In America, Thomas Say and Constantine Rafinesque sent type specimens to museums in London, Paris, and Vienna. Philadelphia lawyer and collector Isaac Lea named around 85 North Carolina mussel taxa from specimens sent to him by state geologist Ebenezer Emmons, among others.

In his zeal, Lea named a few too many species of the common river mussel in the genus Elliptio, leaving today's taxonomists the task of culling the mistakes and redefining relationships among accepted species. Arthur Bogan, State Museum curator of aquatic invertebrates, found 42 names on record for one North Carolina species, Elliptio complanata. Elliptio is a unionid mussel, belonging to the largest family of freshwater bivalves. Outdated nineteenth-century classifications of this important group created confusion among scientists studying unionid populations and their environmental roles. With the aid of DNA analysis, Bogan is working to revise the classification of higher taxa of Unionoida based on evolutionary relationships.

A better understanding of mussel species allows biologists to use early records to track fluctuations in mussel populations over the years. Much of the record shows a decline in mussel species, underscoring how sensitive mussels are to changes in their environment. Mussels do not move around much, and individuals may live a very long time (up to 130 years!) in one place. Because mussels are communal and tend to reside in beds located in shallow streams, changes in water flow created by dams can wipe out entire mussel populations. Silt from construction and agricultural runoff can harm mussels, which are filter feeders. They concentrate pollutants in their tissues, and incorporate metals into their annual shell layers.

The Carolina heelsplitter filtered the waters of Waxhaw and Goose Creeks in Union County long before Isaac Lea first described the species. Environmental problems caused by twentieth-century development and agricultural runoff placed this species on the federal list of endangered species, along with the Tar River spinymussel, dwarf wedge mussel, and little-wing pearlymussel. The North Carolina Wildlife Resources Commission and others are working to conserve the mussels' fragile habitats through landowner agreements.

The mussel's life cycle is so finely tuned to its environment that it depends on the presence of a host fish specific to its species. A female mussel incu-

bates the larvae, called glochidia, for up to a year. Then the pinhead-sized larvae hitch a ride on the gills or fins of a species-specific fish host. There they metamorphose into juvenile mussels.

Certain mussels have developed an array of lures that fool the fish into picking up hitchhiking glochidia. In the current, females release a glochidia-laden packet that looks to the fish like a tasty minnow or worm. When the fish takes the bait, all it ends up with is a mouthful of mussel larvae. A few larvae manage to imbed themselves in the gills of the fish, incubate for one to six weeks, and then burst out of the gills to settle in the stream bottom. Without the host fish stage, the glochidia die. When a dammed river or other environmental stress causes the host fish to move out of an area, entire populations of mussels may die.

Freshwater mussels mean food to a community of riverine animals: fishes, ducks, herons, turtles, muskrats, raccoons, river otters. In a round-about but essential fashion, mussels enhance the oxygen supply in water systems. As a mussel feeds, it filters bacteria, algae, and detritus though its gills, clearing the water. Clear water means deeper sunlight penetration, which increases photosynthesis, which produces oxygen. Any organism that relies on clean water (including humans) benefits from the filtering action of the region's dwindling mussel populations.

Phylum Arthropoda: Insects

Insects, being of great interest to farmers, were the first animal group in North Carolina to receive the full attention of a state-funded biologist. In 1900 state entomologist Franklin Sherman began a statewide survey of insects, listing 1,707 known species for North Carolina. By 1902 his insect collection had grown to 30,000 specimens, all accessible to farmers and the general public at the State Museum. He enlisted C. S. Brimley, then a private collector, to help with field surveys. The two became lifelong friends and coauthored many papers on the state's tiger beetles, butterflies, dragonflies, and grasshoppers, among other invertebrates. A score of amateur insect collectors across the state contributed to the collection, especially the Reverend A. H. Manee of Southern Pines.

After 20 years as a freelance entomologist and collector, C. S. Brimley joined the North Carolina Department of Agriculture to head the state insect survey. By then the number of known insect species had nearly tripled. When Brimley completed *The Insects of North Carolina* in 1938, more than 9,600 species were known in the state. He realized there were more to be found. "What we need to increase our list is not more insects," he wrote, "but more entomologists in the state." Brimley predicted the total number of insect species to be around 20,000—not a bad estimate, according to today's entomologists. The state's collections, which now include series from around the world, are curated by staff entomologists at North Carolina State University.

(*opposite*)
Yellow lance mussel, *Elliptio lanceolata*. From Isaac Lea, *Observations on the Genus* Unio, 1828. Courtesy of Rosamond Purcell

(*right, top*)
Tiger swallowtail, *Papilio glaucus*, is common throughout eastern North America. Drawing by John White, 1584.

(*right, bottom*)
One of the rarest butterflies in the country, Saint Francis' satyr, *Neonympha mitchellii francisci*, is found along one stream in the Sandhills region of the Coastal Plain. Courtesy of Stephan Hall.

C. S. Brimley, photographed with his insect collecting net and telescope, accounted for more than 9,600 insect species in his 1938 book, *The Insects of North Carolina*. Courtesy of William M. Craven.

By far the most successful group of invertebrates is the class Insecta, which includes three-fourths of all known animal species. A 220-million-year-old fossilized beetle, preserved in the fine-grained shale of the Piedmont Triassic basin, inhabited a North Carolina ecosystem that predated flowers. The first animals to fly, insects became pollen gatherers when flowering plants appeared a hundred million years ago. Beetles were among the first to gather meals of pollen. Now one out of every four animal species is a beetle.

Second in number of species only to beetles, the butterfly and moth order, Lepidoptera, includes at least 2,000 species found in North Carolina. A tenth of these are butterflies. One small butterfly, the Saint Francis' satyr, exists in only one location in the world—along a single stream in the Sandhills region of North Carolina. Suitable habitat for this rare butterfly is shrinking. It needs woods kept open by natural fires, and the suppression of fire in the region is a threat to its survival.

The majority of lepidopterans are moths—a huge group of animals with great variation in size, color, and behavior. The Cherokee watched adult moths fly into flames, and reasoned that moths could be used to treat "fire diseases" like sore throat. Many entomologists now believe that this attrac-

tion to light is triggered by a built-in steering mechanism that uses the Moon as a beacon. Bright lights on Earth throw off this navigation system, and a moth's flight pattern veers relentlessly into the light.

The giant silkworm moths of the family Saturniidae are probably the most spectacular insects in the state. The brightly colored cecropia moth may be North Carolina's largest insect, with a wingspan of up to six inches. Pale green luna moths sport a long trailing "tail" on each hind wing. The two large eyespots on the hind wings of the bull's eye moth may scare off predators.

The warm North Carolina climate allows some saturniids to produce two or three generations in one summer. Eggs hatch in a week or so into larvae (caterpillars). Big saturniid caterpillars are often brightly colored or bristly with spines and hairs. They live to eat, feasting on the succulent leaves of trees and bushes for five or six weeks.

After the feast, the caterpillar spins its silken cocoon and becomes a pupa—the stage in which the caterpillar transforms into a moth. On the night after she emerges from the cocoon, the adult female moth emits a chemical that attracts males from as much as six miles away. She then mates and lays her eggs. Adult saturniids live to reproduce, not to eat, so within a week the moth has starved to death. The last brood of the season will over-winter in sturdy cocoons so tough that Native Americans once used them for rattles and containers.

Stages in the life of the silkworm moth played a role in the state's traditions. Fishermen used the silk glands of the caterpillar to make a very strong leader for fishing lines. Silk Hope in Chatham County produced silk for collars before the Civil War, but a blight on mulberry plantations put an end to cultivation of cocoons of Asian silkworm moths used for store-bought silk cloth. North Carolinians once used native silkworm cocoons for homespun silk. In 1879, the North Carolina State Exposition (the early State Fair) offered a prize in the "Ladies' Work" category for "best specimen sewing silk, from native cocoons," probably those of polyphemus or promethea moths. The State Museum displayed homemade silk made by ladies in the Piedmont into the 1940s.

Public perception of the world of invertebrates is changing as research and new technologies make possible a keener appreciation of the social organization, sensory capabilities, and enormous diversity of invertebrate groups. Museum displays of insect pests and oyster harvesting have given way to the Arthropod Zoo, where visitors experience the compound vision of insects, witness the colonial life of harvester ants and termites, and view close-up videography of moth mating rituals and walking stick locomotion. Visitors learn that invertebrates—small in individual size, tremendous in numbers and diversity—play a vital role in a healthy ecosystem, and serve as prime indicators of environmental change.

Fishes

"We are told by the Indians of a great many strange and uncouth shapes and sorts of Fish," wrote John Lawson in his 1709 natural history of North Carolina. A self-described "great Lover of Fish," Lawson counted 19 kinds of freshwater fishes in North Carolina, and believed he had accounted for around one-third of the total species in the state. By 2000, ichthyologists had identified 210 freshwater species and more than 700 saltwater species occurring in the state's waters.

The business of harvesting fish has driven much of the research on fish diversity in North Carolina. A high-ranking federal fisheries biologist, Hugh Smith, counted 345 fish species for the North Carolina Geological and Economic Survey in 1907. In his report, *The Fishes of North Carolina*, Smith attributed the diversity and abundance of North Carolina fishes to the state's geology—its varied terrain; the high number, length, and volume of its streams; and the presence of coastal sounds and estuaries. Tropical saltwater fishes reminiscent of the West Indies and Florida swam offshore, along with cod and tautog from the northern Atlantic. Smith marveled at the "immense schools of mullet, squeteague, menhaden, blue-fish, croaker, and spot" in the sounds and offshore waters, and the unusual diversity of suckers, minnows, sunfishes, and darters in freshwater streams.

Lumped under the all-purpose heading "fishes," three classes of vertebrates share the possession of gills and an often streamlined shape suited for movement in water. This huge assemblage of animals attracted the study of

"Many species of fishes were first made known from North Carolina waters. . . . Other species here exist in greater abundance than in other states" (Smith, *The Fishes of North Carolina*, 1907).

Florida pompano, *Trachinotus carolinus*, is one of hundreds of fish species found in North Carolina waters. Courtesy of Rosamond Purcell.

some of the most prominent Victorian-era naturalists: Georges Cuvier in France, Louis Agassiz at Harvard, and Edward D. Cope at the Academy of Sciences in Philadelphia. Investigations conducted by these men into relationships among living and fossil fishes laid the foundations for ichthyology and paleontology.

Traditionally, taxonomists list first the most ancient fish lineages. The class Agnatha, the jawless fishes, arose in the Silurian period around 438 million years ago. Living agnathans inhabit North Carolina waters today: lampreys are found in freshwater streams in the mountain and coastal regions, sea lampreys return from salt water to spawn in freshwater streams, and hagfishes inhabit offshore waters. These primitive vertebrates can exasperate fishermen. John Lawson obtained a single lamprey specimen that was caught in a weir "by the Indians, who would not eat him." Today they are still spurned as food in North America, though considered acceptable in several Eurasian cuisines. The saltwater hagfish, considered by some to be a more primitive, sister group to the vertebrates, lives in cold, deep waters at the edge of the continental shelf. It eludes its predators by clogging their gills with the slime that oozes from dozens of pores on its two-foot-long body.

Two classes of jawed fishes emerged in the late Devonian period around 370 million years ago: fishes with cartilage only, Chondrichthyes, and the bony fishes, Osteichthyes. All major groups of Chondrichthyes—the sharks, rays, skates, and ratfishes—inhabit the state's coastal waters. Rays of "monstrous Size and Strength" were known to eighteenth-century English settlers as "divel-fish." John White included a hammerhead shark in his sixteenth-century sketches of the fishes of Carolina. An abundance of offshore sharks supported a sharkskin shoe leather tannery in Beaufort during World War I.

The size of sharks, along with their fearsome reputation, impressed visitors to the early State Museum. On display were a rare 16-foot thresher shark netted off Wrightsville Beach and a 35-foot-long, five-ton whale shark that beached in the lower Cape Fear River in 1934, the first record north of Florida for this largest of living fishes. Director H. H. Brimley shared the public's fascination with sharks. His 1935 essay, "Sharks I Have Known," begins: "You'd better not be too familiar, Or scratch him too much on the head / And, perhaps, when it comes to a showdown / He is safer—and better—when dead." Of his two showpiece shark specimens he wrote, "Both the Whale Shark and the Basking Shark—the latter reaching a length of about forty feet—are great, harmless, sluggish creatures, the teeth being both very small and very numerous and in no way suited for inflicting dangerous wounds."

Biologists believe that all four-limbed vertebrates evolved from Osteichthyes, the bony fishes, which are more closely related to humans than to sharks. The most diverse class of vertebrates, bony fishes account for a good share of North Carolina's biodiversity. They inhabit the waters from saltwater sounds and estuaries to freshwater lakes, streams, and wetlands in 17 river basins.

(opposite)
Blackbelly rosefish, *Helicolenus dactylopterus*. Courtesy of Rosamond Purcell.

(*right, top*)
Saltwater fishes near Roanoke Island include a hammerhead shark. Drawing by John White, 1584.

(*right, bottom*)
Fishes from the collection of Edward D. Cope. Left to right, top to bottom: northern studfish, banded darter, blacktip jumprock, Tennessee snubnose darter, saffron shiner, and tangerine darter. The banded darter, saffron shiner, and tangerine darter are found in western river basins of North Carolina. From *Journal of the Academy of Natural Sciences* 6, Part 3 (1869): Plate 24.

(*opposite*)
Scalloped hammerhead shark, *Sphyrna lewini*. Courtesy of Rosamond Purcell.

Five mountain basins on the western side of the eastern continental divide, which runs through North Carolina's mountains, drain to the Gulf of Mexico. The twelve basins to the east of the divide drain to the Atlantic Ocean. While a few fish species inhabit both eastern and western streams, the fishes on either side of the divide represent distinct faunas. Such diversity attracted the attention of early ichthyologists. On an 1869 trip to the state, Edward Cope found the number of fish species "rather greater in [North Carolina] than in Virginia; the mountains here constituting a much more important topographical feature."

This separation of fish faunas at the eastern continental divide inspired an insight into evolutionary theory by one of the foremost fish men of the Victorian era, David Starr Jordan. World traveler, poet, and self-styled "minor prophet of democracy," Jordan collected folk songs and pitched in local baseball games while collecting fishes in the mountains of North Carolina. He noted that in waters on either side of the divide, "the same general types pre-

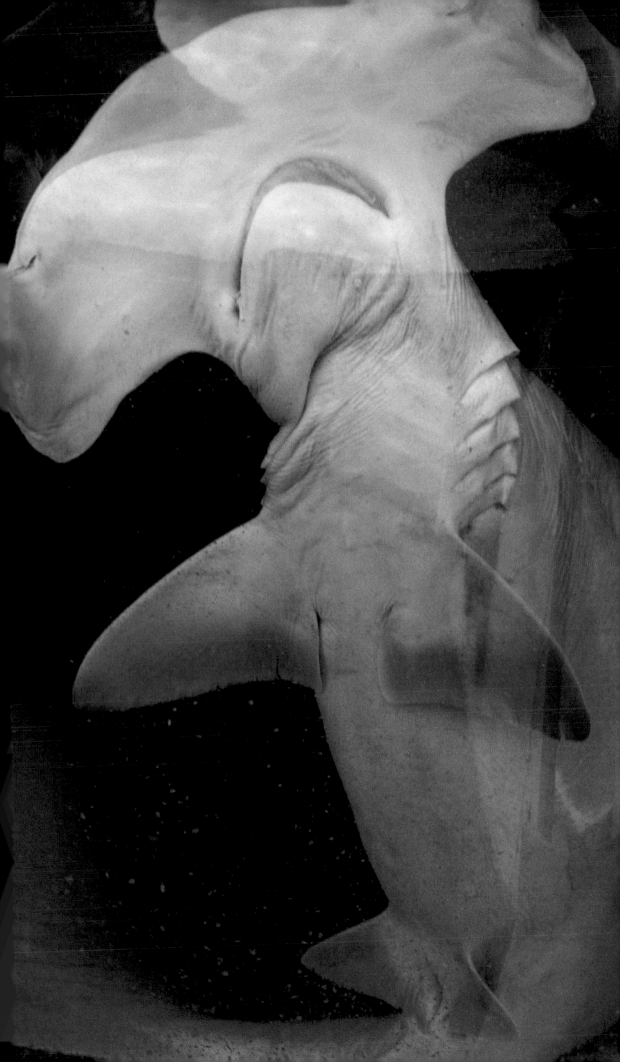

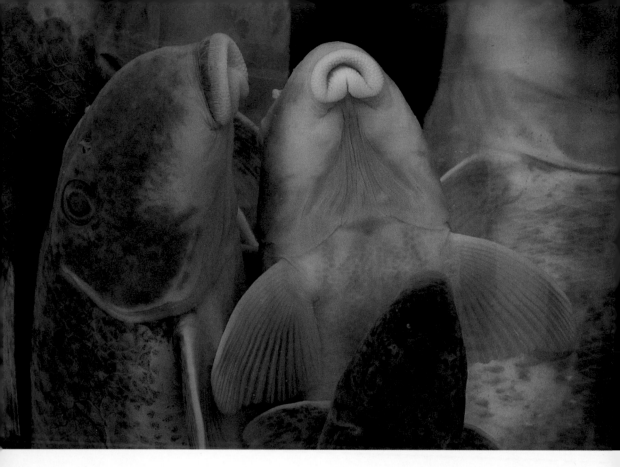

Northern hog sucker, *Hypentelium nigricans*. The northern hog sucker probably originated in the Mississippi River basin, and later entered the eastern Roanoke River basin. It is now found in both eastern and western streams in North Carolina. Courtesy of Rosamond Purcell.

vail, but among the small, non-migratory fishes—minnows and darters especially—the species are different, each river having its own kinds, with their nearest relative occurring in the next, not in the same, stream." Jordan, once a student of the staunch creationist Louis Agassiz, surmised that related mountain species arose from a widespread ancestral species whose populations had been isolated by a physical barrier, the barrier in this case being the continental divide. His "Jordan's Law" became the basic tenet of vicariance biogeography, an area of study that flowered a hundred years after Jordan made his observations.

Some species probably arose in the west but are now found in both the east and west. Biologists believe that they crossed the divide by a geologic process called stream capture. Streams on the eastern side of the divide gradually eroded surrounding rock and intercepted parts of western streams, thus acquiring the fish of the western drainage. The northern hog sucker, a tube-shaped fish with big lips, likely evolved in the Mississippi River drainage. Streams of the Roanoke River basin captured western streams inhabited by the species, and the northern hog sucker became a fish of the Roanoke. It shares the river with the Roanoke hog sucker, probably an earlier offshoot of an ancestral species common to both hog suckers.

In the geologically stable Appalachian range, fishes evolved into many species with an amazing degree of adaptation to their environments. The small, well-camouflaged mottled sculpin lives in clean, clear mountain streams. Aided by wide pectoral fins and a large mouth, it crawls along gravel creek beds ready to ambush insect larvae and larger food like crayfishes and small fishes. The streamlined shape of the longnose dace gives that fish stability in the swiftest streams as it forages for mayfly and midge larvae.

North Carolina has six freshwater fishes found in no other state; 31 species are found only in North Carolina plus Virginia or South Carolina. In the shallow waters of Lake Waccamaw in the southeastern corner of the state, three common stream fish—a killifish, a silverside, and a darter—evolved into distinct species adapted to lake life. Two shiner species arose in North Carolina's Piedmont streams. The Cape Fear shiner occurs only in a few rivers and streams in the Cape Fear River basin, and the pinewoods shiner is found only in the Tar and Neuse river basins. The Sandhills chub and pinewoods darter are not found outside the distinctive south-central Sandhills ecosystem straddling the border of North and South Carolina.

The Carolina madtom, a small, handsomely patterned, secretive catfish found only in the Neuse and Tar river basins, earned the name *Noturus furiosis* from David Starr Jordan, who discovered it near Raleigh. Jordan wrote that its "special organ of offense," a venomous pectoral spine, could inflict wounds on the unwary fish collector that are "exceedingly painful, like those made by the sting of a wasp."

North Carolina claims four species of primitive bony fishes, survivors of ancient orders that were once widespread. On his 1869 collecting trip to the state, Edward Cope recorded paddlefish, bowfin, gar, and sturgeon. In the cold waters of the western French Broad River in the Tennessee River basin, he found the only North Carolina specimen ever recorded of the paddlefish. One of the strangest of freshwater fishes, the six-foot-long paddlefish is built like a paddle on the front end, and has a distinctly sharklike tail. Only two species of paddlefish exist today; the paddlefish of North America and a huge Chinese species that grows to seven meters long. Paddlefish are still locally common in the Midwest and Great Lakes regions.

Cope published reports of the bowfin from the Neuse River basin. The only living species of an ancient order, the bowfin is found throughout the Coastal Plain in ditches, canals, and slow-moving creeks. Eastern fishermen know the three-foot-long fish as grinnel, blackfish, or mudfish. Lawson spurned it in 1709: "Grindals [grinnels] are a long scaled fish with small eyes . . . they are a soft sorry fish, and good for nothing."

The longnose gar is a large, ancient fish found throughout the state's eastern river basins and in the mountains in larger tributaries of the Tennessee River. Skinned and boiled, the white flesh of the gar was highly prized among coastal African-Americans in the late 1800s, when a unique gar fishery operated in New Bern. In Edenton, the species was known as doctor-fish; John Lawson reported, "The gall of this fish is green, and a violent cathartick, if taken inwardly." The armor-like skin of the gar was used to cover boxes, sword hilts, and even wooden plow-shares. The gar thrives in warm, low-

Atlantic sturgeon, *Acipenser oxyrinchus*. Once important as a prime producer of caviar from North Carolina, the Atlantic sturgeon is an ancient migratory species that can grow to 12 feet long and 500 pounds. Atlantic sturgeon was overharvested in the nineteenth century, and is now listed as a species of special concern in the state. From Smith, *The Fishes of North Carolina*, 1907.

oxygen waters, and gulps air at the surface, filling a special air sac that functions like a primitive lung.

In 1588, Sir Walter Raleigh's expedition scientist Thomas Hariot was the first to record the state's abundance of sturgeon in the spring, when the fish leave salt water and return to their natal rivers to spawn. By the eighteenth century, John Lawson said of sturgeon, "we have plenty, all the fresh Parts of our Rivers being well-stor'd." By the 1800s, the sturgeon became the most valuable fish per pound in North Carolina, prized for its flesh and its eggs, which were treated in brine and sold as caviar.

But excessive fishing decimated sturgeon populations in the 1880s and 1890s. Fisheries biologist Hugh Smith urged the state to restore the sturgeon population "to something like its original abundance, if this is now possible," by issuing a moratorium on sturgeon fishing, and by stocking sturgeon in coastal rivers. The Atlantic sturgeon, an ancient species that has survived for 200 million years, is now listed as a species of special concern in the state. Though the fish can live more than 50 years and attain 12 feet in length and 500 pounds, large individuals are rarely found today. A smaller, endangered species, the short-nosed sturgeon, virtually disappeared from the state's catch for decades. In 1998 biologists captured the second shortnose sturgeon known from the Albemarle Sound since 1881.

Anadromous fishes like the sturgeon live their adult lives in the ocean, migrating up Coastal Plain rivers in the spring to spawn. When rivers ran free and clear prior to European settlement, Native Americans counted on fish feasts during spring spawning season. Thomas Hariot wrote in 1588, "For foure monthes of the yeere, February, March, Aprill and May, there are plentie of Sturgeons: And also in the same monethes of Herrings . . . both these kindes of fishe in those monethes are most plentifull, and in best season, which wee founde to bee most delicate and pleasaunt meate." The herring (now recognized as two species, the blueback herring and alewife) still creates excitement among fishermen in tributaries of Albemarle and Pamlico Sounds when they return to their home rivers to spawn. River herring topped the list of migratory food fishes in the nineteenth century, and for many years North Carolina caught the most of any state.

Shad, *Alosa sapidis-sima* (*left, top*); work-ers sort the catch at the Capehart shad fishery on the Cho-wan River in the 1880s (*left, middle*). When shad migrate upstream to spawn in the spring, some North Carolina towns hold shad run cele-brations. Drawing from Smith, *The Fishes of North Caro-lina*, 1907; photo-graph from North Carolina State Mu-seum Archives.

(*bottom*)
The U.S. Bureau of Fisheries built a state-of-the-art shad hatchery, com-plete with automatic hatching jars, near Edenton in 1900 to supply shad stock for nearby waters. From Smith, *The Fishes of North Carolina*, 1907.

Shad runs are celebrated by festivals and feasting in eastern North Caro-lina. Shad is served up baked, fried, smoked, salted, and pickled, accompa-nied by the highly prized roe. Members of the herring family, the American shad and hickory shad once spawned in great numbers from Albemarle Sound to the Cape Fear River.

When the shad catch declined in the late nineteenth century, the U.S. Fish

Commissioner sent the steamer *Fish-Hawk*, a floating shad hatchery, on an annual mission to seed Albemarle Sound with young shad hatched from eggs that would otherwise have been sent to market. The pride of the Fisheries Commission was a permanent hatchery built near Edenton in 1900, designed to plant millions of young shad in the sounds—so many that "the perpetuity of the run would be insured." But the hatchery was severely limited by the lack of spawning fish, which were depleted by fishermen before they could ascend the rivers in spring. The 1904 state legislature passed laws safeguarding the spawning fish, with some success.

Over the course of the next century, the cumulative effect of dams, overfishing, and stream channelization cut the state's shad catch by 95 percent. Dams and locks built earlier in the century blocked the Cape Fear River shad run. Eventually the U.S. Army Corps of Engineers opened the locks in the river to create a flow of water during spawning season. Enough shad reached their traditional spawning grounds to build a small but viable population once again. The Quaker Neck dam on the Neuse River at Goldsboro cut off hundreds of miles of breeding habitat for shad, herring, striped bass, and other migratory fish when it was built in 1952. The shad catch fell from 700,000 pounds to 25,000 pounds before the dam's owner, CP&L, and the North Carolina Department of Environment and Natural Resources demolished it in the 1990s, allowing the fish to return to traditional spawning grounds.

Spawning behavior can be highly specialized, especially when fish species compete for space in limited areas. The fieryblack shiner, a small species found only in the Carolinas, spawns several times a year from June to September. The female rushes through a crevice guarded by the male and deposits her eggs. The male follows to fertilize them, and resumes his guard duty. The bull chub, endemic to North Carolina and Virginia, is also called horny head, for the cranial bumps it acquires in breeding season. The male bull chub makes a large nest of mounded rocks for its spawn. As the female releases her eggs, they are fertilized by the male and drop safely into gaps between the stones. The male allows smaller species of fish to share his nest; most of these sport bright red breeding colors, creating a tropical scene in Piedmont streams.

Five species of the sucker family spawn in one river on a time-share plan, using the spawning grounds in succession from early to late spring. Found in every freshwater fish habitat in the state, from mountain streams to lowland creeks, suckers spawn like clockwork in spring. Old-timers claim, "When the redbuds are blooming, the suckers are spawning."

Native Americans fished a species of sucker, the robust redhorse, from the Yadkin River when it spawned in April and May. In a great migration upstream, males arrived first, sparring with other males for territory. Females followed swollen with eggs, swimming in water so shallow their backs were exposed. The large fish were easy to harvest by stick-fishing.

Edward Cope first described the species when he collected the fish in the Yadkin River in 1870. It is one of a few suckers in the world that has special molar teeth on its gill arches capable of feeding on mollusks. Cope decried the widespread method of catching the spawning redhorse in efficient spring-nets. "Too many of the people with the improvidence characteristic of ignorance, erect traps, for the purpose of taking the fishes as they ascend the rivers in the spring to deposit their spawn. Cartloads have thus often been caught at once." Cope believed the state of North Carolina should make it a policy to destroy "spring traps and weirs every spring, at the time of running of the fishes, to allow the escape of immense numbers of them, before the traps could be repaired."

Cope's description was the first and last notice of the robust redhorse in the scientific literature for 122 years. Its name was inadvertently transferred to a smaller, very common sucker species. Other specimens may have been collected in that stretch of time, but were unidentified or misidentified. Finally in 1985 a half-rotted specimen from the Pee Dee River was identified from Cope's description, and the misuse of the name was recognized by taxonomists. A second specimen, captured and released by State Museum curator of fishes, Wayne Starnes, and other biologists in the spring of 2000, was a healthy six-year-old individual.

Already declining by 1870, the robust redhorse has been pushed to the edge by overfishing, deforestation, damming of rivers, and predation from introduced species. The predatory flathead catfish, introduced in Piedmont rivers as a game fish, has taken its toll. Though a small robust redhorse population exists in Georgia, most of its members are more than 15 years old. With little spawning activity and young coming along, the species is close to extinction. Several agencies are working to improve the future of the robust redhorse through habitat management and close monitoring of existing populations.

Since the 1870s, wildlife managers have intentionally introduced exotic fish species into native waters, despite the disruption the practice has caused in native populations. The U.S. Fisheries Commission introduced the Eurasian carp in many states during Reconstruction, to be cultivated as a food fish. The carp is now a permanent resident across the nation. The rainbow trout, a native of western North America, was first stocked in Blue Ridge mountain streams in 1880. Rainbow and northern brook trout pushed native brook trout populations far upstream, restricting their territory. Stocking of popular freshwater game fish species, from striped bass to bluegill, has dramatically altered fish populations in lakes and streams across the state.

Freshwater fish populations may fluctuate dramatically because of changes in the environment (river damming, stream channelization, siltation from development, and pollution), from overfishing, and from attack by introduced predators and parasites. The freshwater fish collection of the State Museum helps biologists track the rise and fall of fish populations over

Brook trout, *Salvilinus fontinalis* (*top*); rainbow trout, *Oncorhyncus mykiss* (*middle*); brown trout, *Salmo trutta* (*bottom*). Courtesy of Duane Raver.

(*opposite*) Museum director H. H. Brimley, shown with largemouth bass mount, was an ardent promoter of recreational fishing in the state. Fish species introduced for sport fishing like the largemouth bass have altered native populations in the state's streams and lakes. Rainbow trout from western North America readily adapted to mountain streams, crowding native brook trout upstream. North Carolina State Museum Archives.

decades—an important indicator of environmental health. The first fish distribution guide for the continent, the *Atlas of North American Freshwater Fishes*, was published by the State Museum in 1980.

North Carolina's saltwater fishes inhabit 2,500 square miles of sounds, bays, marshlands, and tidal creeks along 320 miles of coastline. Around 90 percent of the fish caught off the coast spend part of their life in estuaries, where young animals feed on algae and detritus and find shelter in beds of seagrass.

The productive habitats of the continental shelf extend eastward into the Atlantic Ocean for 40 miles or more. Fishes of the subtropics and the North Atlantic meet here where the powerful Gulf Stream mixes with the Labrador current. Beyond that, the plunging continental slope harbors one of the richest deepwater slope habitats on Earth.

In the hardbottom habitats of the continental shelf, large fish like greater amberjacks prey on squid and schools of smaller fish. Bathed by occasional

offshoots of the Gulf Stream current, the hardbottom is home to tropical and temperate species alike, from angelfish and moray eel to tautog and black sea bass. More than 300 fish species find shelter or hunt for food among the rocky limestone outcrops.

These fish inhabit underwater landscapes populated by land animals when the continental shelf was above sea level in the last ice age. River valleys cut into the sedimentary rock of the shelf. As glaciers began to melt around 10,000 years ago, sea level rose. Ocean waters flooded the shelf and its river valleys, and marine life moved in.

Beyond the shelf, where converging currents collide, nutrients well up to the surface, feeding a bloom of phytoplankton that sustains an enormous open-ocean food web. Larger fish, birds, and other animals gather near the surface to feed. The fallen remains of these organisms create blizzards of organic sediments known as marine snow, which feeds a deepwater food web on the steep continental slope below. Fishes on North Carolina's continental slope are up to ten times as abundant as fishes living on most of the world's slope habitats. The virtually lightless habitats of the slope have been underwater for 200 million years, and are home to bioluminescent creatures with remarkable adaptations to their high-pressure, dark environment.

Further offshore is the Sargasso Sea, a vast area extending into the central Atlantic where floating rafts of sargassum seaweed provide homes to millions of organisms, some fantastically camouflaged in their weedy habitat. Looking more like seaweed than fish, the sargassumfish prowls these jungles with limb-like fins. A complex food web links the organisms of this diverse community, from tiny invertebrates to dolphin fish, bridled terns, sea turtles, and porpoises.

Thomas Hariot touted the bounty of fishes he found at Roanoke Island in 1588. "There are Troutes, Porpoises, Rayes, oldwives [alewives, herring],

Sargassumfish, *Histrio histrio*. The sargassumfish virtually blends in with its habitat—wide mats of sargassum seaweed found off the coast of North Carolina and into the central Atlantic Ocean. Courtesy of David Lee.

Mullets, Plaice [flounder], and very many other sortes of excellent good fish, which we have taken & eaten." Now, flounder, menhaden, tuna, weakfish (sea trout), croaker, and mullet lead the saltwater catch. In the 1890s the catch was worth a million dollars. By 1907 more fisheries operated along the North Carolina coast than in any other state.

At the end of the nineteenth century, striped mullet was the mainstay of the state's commercial fisheries. Naturalist H. C. Yarrow claimed mullet was "the most abundant [fish] of the locality, and affords sustenance and employment to thousands of persons on the coast of North Carolina . . . the numbers taken are simply enormous." Biologist Robert Coker claimed the mullet was so plentiful that "the loss of no other fish could so embarrass the fisherman as a failure of mullets."

To promote North Carolina fisheries at the 1893 World's Columbian Exposition in Chicago, H. H. Brimley re-created a mullet hut used by fishermen on the Outer Banks. Moss-draped palmetto trees shaded the hut, and fishing and oystering implements stood at the ready amid oyster, clam, and other seashells shipped from the coast. Mounted fish, overhead nets, and fishing traps rounded out the scene. Enthusiastic visitors to the exhibit claimed this tribute to mullet fishing was second only to the elaborate displays of Japan, Canada, and Norway. When not on the road, the fish mounts went on display at the State Museum.

Interest in collecting saltwater fishes increased as the state catch grew. Several of the big names in the young science of ichthyology traveled to Beaufort, among them Theodore Gill, Army doctors H. C. Yarrow and Elliott Coues, and the ubiquitous David Starr Jordan, who spent a delightful month in 1878 at the "mildly fashionable summer resort" collecting fishes in the surrounding sounds.

Fish biologist H. V. Wilson found that the geology of the coast created ideal conditions for the collector: "The coast configuration forms a remark-

able natural trap into which fish, migrating northwards, fall. It is doubtful whether a better place can be found anywhere on our coasts for the carrying out of observations on oceanic species and on bay and river species during the oceanic period of their life. The seining . . . at Cape Lookout has been extremely interesting and successful, in variety of forms and the number of individuals taken."

Honeycomb moray eel, *Gymnthorax saxicola*. Courtesy of Rosamond Purcell.

Recognizing the economic importance of North Carolina's saltwater fishes, the U.S. Bureau of Fisheries built a full-scale research lab at Beaufort Harbor in 1902. The steamer *Fish-Hawk* made offshore collecting runs, recording many new fish species for North Carolina waters. Fishermen, fish dealers, and Coast Guard crew sent unusual specimens to the fisheries lab, which were sometimes forwarded to the collection at the State Museum, including a rare sharptail mola caught in 1909.

The state's first recorded specimen of the huge ocean sunfish, *Mola mola*, was found by William Haskill at the Cape Lookout Life Saving Station in 1889. Looking like a giant fish head with no body or tail, *Mola mola* is often mistaken for a shark in offshore waters because its large dorsal fin may project above the surface. Individuals usually weigh 200 to 400 pounds, though the record weight for the species is over two tons. In 1904 James Willis picked up a small *Mola mola* on the beach at Cape Lookout and sent it to the Beaufort fisheries lab. Later that spring in the same area, fishermen

H. H. Brimley re-created an Outer Banks mullet shack at the 1893 World's Columbian Exposition in Chicago to draw attention to the state's growing commercial fishing industry. From Smith, *The Fishes of North Carolina*, 1907.

A papier-mâché model of an ocean sunfish, *Mola mola* (*above, left*), receives the finishing touch from a curator before display in the State Museum. Fishermen brought the 1,200-pound fish (*above, right*) into Swansboro in 1926, where H. H. Brimley made notes on the dimensions of the giant fish. North Carolina State Museum Archives.

harpooned an eight-foot, half-ton specimen. Always on alert for large fish for the State Museum, H. H. Brimley and Harry Davis hurried to Swansboro in 1926 to investigate a 1,200-pound *Mola mola* captured in Bogue Inlet by Capt. W. E. Mattocks. From field notes taken at dockside, Brimley created a papier-mâché model of the wide-ranging fish. For 70 years, the huge model, with its large glass eye and tiny mouth built for jellyfish-eating, impressed visitors to the State Museum.

A young fish collector from Farmville mounted many specimens for the State Museum during the Depression years. Roxie Collie made headlines as the only female professional taxidermist in the state and the first woman to serve on the state's curatorial staff. Collie prepared specimens in the winter; in summer she collected fishes in Core and Bogue sounds using a small gas-powered boat, working out of the Beaufort fisheries lab. Her fish mounts enhanced the popular hall of fishes at the State Museum, where mounts of greater amberjack, blue marlin, basking shark, red drum, and a model of the sharptail mola hung suspended above many cases of smaller fishes.

Today the rich habitats of the estuaries and continental slope are the research focus of scientists in the waterside laboratories of the North Carolina Division of Marine Fisheries, North Carolina State University, University of North Carolina at Chapel Hill, Duke University, and the NOAA/National Marine Fisheries Service (the descendant of the 1902 fisheries lab). A collection of 380,000 marine fish specimens assembled by Frank J. Schwartz of UNC's Institute of Marine Sciences was transferred to the State Museum in 1996. His 1,500-gallon collecting jars (actually steel tanks) hold an intact basking shark nearly 15 feet long and a mammoth ocean sunfish. Wayne Starnes called Schwartz's collection "the primary documentation of marine biodiversity for the N.C. region through a good part of this century."

Since Thomas Hariot's survey in 1588, scientists have recognized the extraordinary abundance and species diversity of North Carolina's fish fauna. The State Museum's salt and freshwater fish collection of over a million specimens helps biologists track changes in the state's fish populations— changes that can affect fisheries and recreational fishing, and indicate environmental hazards. Beyond their value as food and sport, live fishes fascinate the thousands who visit the state's three aquaria and the State Museum. Living habitat displays of native fishes allow everyone a glimpse of the diversity of shapes, sizes, colors, and behaviors that have attracted fish collectors to the state for centuries.

An ocean sunfish and sharptail mola hover over the State Museum's Fish Hall, ca. 1950. North Carolina State Museum Archives.

Reptiles and Amphibians

Amphibians and reptiles tend to cluster by geographical region in North Carolina. The Mountain region claims the most diverse group of salamanders in the world, while many of the state's frog and snake species are found in the Coastal Plain. Turtles abound in the Piedmont and Coastal Plain, and the state's five largest turtles inhabit the ocean and estuaries. The American alligator, a crocodilian, lives in coastal freshwater creeks, swamps, and marshes south of Albemarle Sound.

All told, North Carolina has around 58 salamanders (herpetologists disagree on the exact number of species), 30 frogs, 21 turtles, 12 lizards, 37 snakes, and 1 crocodilian. Six of the state's snakes are venomous. The reclusive coral snake has the most toxic venom, but no records exist of coral snake bites in North Carolina. Considered most dangerous to humans is the eastern diamondback rattlesnake, a very rare snake of the southeastern Coastal Plain.

Although taxonomically as different as birds are to mammals, reptiles and amphibians are often grouped together. Naturalists have struggled for centuries to classify these animals. In the early 1700s John Lawson listed among North Carolina's "Insects" the alligator, snakes, lizards, frogs, turtles, "eel-snake," and "wood-worm." In 1755, Prussian collector Jacob Klein coined the term "herpetology," from the Greek *herpeton*, (crawling things), for the study of all limbless animals, including worms (and excluding frogs, turtles, and lizards). The renowned taxonomist Linnaeus (who abhorred snakes and

"They might as well have call'd it a Glass-Snake, for It is as brittle as a Tobacco-Pipe" (John Lawson, *A New Voyage to Carolina*, 1709).

Eastern glass lizard skeleton. Courtesy of Rosamond Purcell.

Eastern diamondback rattlesnake, *Crotalus adamanteus*. Courtesy of Alvin Braswell.

frogs) grouped the amphibians, reptiles, and cartilaginous fishes in the class Amphibia in his great classification scheme of 1758. In 1825, P. A. Latreille placed amphibians and reptiles in separate classes, where they remain today. Herpetology embraces both classes, and students of amphibians and reptiles call them "herps," for short.

Herps are ancient animals, by and large. Around 360 million years ago, the ancestors of modern amphibians were among the first vertebrates to step onto dry land. Still never too far from water, amphibians must always keep their skin moist. Reptiles reached their peak of diversity in the Mesozoic era (230 to 65 million years ago). Their dominance on Earth ended with that era, when dinosaurs and 80 percent of other reptiles became extinct. Of the modern reptiles found in North Carolina, turtles and alligators have probably been around the longest. Turtles that we would recognize lived 150 million years ago, alongside dinosaurs and giant insects. In the Coastal Plain, turtle fossils from the Cretaceous period (144 to 65 million years ago) turn up regularly.

Salamander diversity has made the southern Appalachian Mountains one of the prime spots in the world to look for clues to evolutionary processes. The region supports the most diverse population of salamander species in the world—58 species in 7 families. Why are so many salamanders here, and what can they tell us about the history of life?

Because the southern Appalachians have remained relatively stable for thousands of years—never glaciated, never a desert—salamanders here have had a long-term lease on habitat suited to their requirements: a temperate climate, high rainfall, and natural communities as varied as spruce-fir forest and mountain bog. As rivers gradually cut deep valleys and gorges around the tall peaks of the region, salamander species were unable to cross from one area to the next. Populations "trapped" on isolated peaks evolved (changed) to become highly adapted to their particular environments. Mountain salamanders became ecological specialists, with exacting requirements for type of habitat, food, or environmental conditions.

The Yonahlossee salamander, a woodland species that favors mossy, damp hillsides, is one of several species found nowhere but the southern Appalachians. Weller's salamander is a small species found only among the leaf litter, rocks, and logs of the highest forests. A subspecies, *Plethodon welleri welleri*, lives only on the top slopes of Grandfather Mountain. The huge hellbender lives under flat rocks in mountain streams. Though fearsome-looking and sometimes more than two feet long, it is harmful only to crayfishes and the small invertebrates on which it thrives. The streamside-dwelling Blue Ridge dusky salamander is part of a species complex that baffled taxonomists until genetic analysis revealed relationships that weren't apparent from external markings.

Jordan's salamander is a single species whose widespread populations look like different species to the casual observer. Presumably widespread in the mountains two million years ago, Jordan's salamander populations were isolated in different mountain ranges. Their descendants are distinguished by colorful markings: those with red cheeks live only in the Great Smoky Mountains; those with red legs are found only in the Nantahala Mountains; and those with no red at all are found over a much greater area. Herpetologists continue to sort out the fine line between species, subspecies, and variations among mountain salamanders.

Three North Carolina turtles—a diamondback terrapin, box turtle, and loggerhead—were the first North American turtles to be documented visually. Naturalist John White recorded the animals in a series of watercolors he painted from 1585 to 1587. The box turtle, White wrote, was "A land Tort which the Savages esteeme above all other Torts." Generations later, colonials raved about the "esteemed" meat of the diamondback terrapin, which was so plentiful that plantation slaves complained about having it as a steady diet.

The diamondback terrapin lives in the coastal marshes and creeks of North America from Cape Cod to the Gulf Coast. In North Carolina, terrapin and settler coexisted until after the Civil War, when a taste for southern terrapin soup took hold in the North. Hunters captured terrapins in the wild and shipped them to holding pens up North, where individuals sold for up to $1 an inch to chefs in East Coast hotels. A willing partner in the business, the State Museum promoted terrapins as an economic resource. Museum curator H. H. Brimley took an "attractive feature" exhibit of terrapins to the 1893 World's Columbian Exposition in Chicago, showing the world that "they can be successfully and profitably grown on our eastern coast." However, by 1900 the terrapin was nearly hunted to extinction. Biologist Robert Coker recommended legislative protection for the turtle without much success. What saved the terrapin was a change in tastes—terrapin soup simply lost its allure in the marketplace, prices declined, and the terrapin population slowly rebounded.

Coker tracked the fortunes of the diamondback terrapin into the 1930s. He

Blue Ridge dusky salamander, *Desmognathus orestes* (*above*); hellbender, *Cryptobranchus alleganiensis* (*right*); Jordan's salamander, *Plethodon jordani* (*bottom*). Around 58 species of salamanders in seven families live in the southern Appalachians. Salamanders have adapted to a diversity of microclimates in the mountains, including mossy hillsides, mountain streams, rock crevices, and the leaf litter of high forests. Courtesy of Alvin Braswell and David Lee.

realized that the terrapin harvest was devastating because of its toll on adult females, which are larger than males and were worth twice as much in the market. Adults live 40 years or more, and females mature at around age five. The female lays 4 to 12 eggs, which are easy prey for raccoons and otters. Gulls and crows feed on terrapin hatchlings. This high loss of turtle eggs and hatchlings is a problem only if females fail to produce eggs year after year. But take enough females out of a population, and overall numbers decline. When hunters left the females alone, the numbers of terrapins gradually increased. However, coastal development in the last quarter of the twentieth century has drained some of the terrapin's salt marsh habitat, leaving adults with less territory in which to breed, nest, and forage. The terrapin is now listed by both the state and federal governments as a species of special concern.

Reptiles generally lead long lives, mature later in life than many other animals, and produce fewer offspring. A bog turtle may live 50 years and lay

(top)
Box turtle, *Terrapene carolina*. John White painted three turtles from North Carolina—a diamondback terrapin, box turtle, and loggerhead. White wrote that the box turtle was esteemed by native Americans. The diamondback terrapin was a staple food of colonists. Drawing by John White, 1584.

(bottom)
Diamondback terrapin hatchling, *Maclemys terrapin*. Courtesy of David Lee.

three to six eggs in a season. Though a raccoon or a mink can gobble those
eggs in a flash, over many seasons the bog turtle can be fruitful and multiply.
The box turtle, North Carolina's official state reptile, is still fairly common.
Though its nest is often raided by foxes, raccoons, and snakes, it replaces itself in the course of a very long life (up to 100 years). Long-lived populations
can handle some loss of eggs to predators, because adults will continue to
produce eggs year after year. But hazards created by humans—hunting,
roads, loss of habitat—remove adults from the population, with historically
drastic results for reptiles like diamondback terrapins and alligators.

The state's alligator story parallels its terrapin history, except that legislation came to the aid of the alligator. John White painted the "Allagatto," then
a common resident in the freshwater creeks and marshes of Roanoke Island,
whose range extended into Virginia. In his 1709 account of North Carolina,
John Lawson reported, "Some people have eaten [alligator tail], and say, it is
delicate Meat," while hunters frequently used the teeth to make chargers for
guns. White alligator teeth, he thought, "would make pretty Snuff Boxes, if
wrought by an Artist."

More valued for its hide than its meat or teeth, the alligator was hunted to
the brink of extinction by the 1960s, and was among the first species placed
on the federal list of endangered species. Legal protection allowed alligator
populations to recover along the southeastern coast, but their range now extends no farther north than Albemarle Sound.

North Carolina alligators are smaller than those in states to the south.
Though John Lawson claimed that some eighteenth-century alligators ex-
ceeded 17 feet in length, the largest on record in the state is a Carteret County
alligator measuring about 12 feet, 4 inches. Males mature at 14 to 16 years
old, and females not until 18 or 19 years old. Hatchlings may stay together for
years.

H. H. Brimley was fascinated by these giant reptiles. In 1926 he collected a group of 24 young alligators from a creek in Onslow County and took them with their seven-foot-wide nest to the State Museum for live exhibit. Always on the lookout for the big 'gator, Brimley pursued the reptiles through coastal swamps for many years. One Craven County specimen he collected in 1905 was on display at the State Museum for 95 years, where generations ventured to touch its hard scaly hide. The *News & Observer* reported the hunt: "The grandfather of all the 'gators was seen and a bullet sent his way to remind him that he, too, was needed [for the Museum]. . . . Curator Brimley says that he will never be content until the monster also adorns the Museum collections." In 1927, a hunter sent to the State Museum the remains of an 11½-foot-long, 500-pound alligator from Onslow County. Brimley's mount of this alligator's skull remained on exhibit for fifty years.

At the turn of the twentieth century, H. H. Brimley foresaw the conditions that would lead to a new crisis for the reptiles of the coast. Of his favorite collecting grounds on the Outer Banks he wrote, "Someday Cape Hatteras will be a great resort—but not in my day I hope." Commercial development now skirts the edges of Nags Head Woods, one of the last remaining examples of mid-Atlantic maritime forest. Sheltered by ancient forested dunes, a remarkable variety of habitats exists here: grassland, scrub, live oak maritime forest with 500-year-old trees, pine flatwoods, and freshwater ponds.

The area is an outpost for reptiles and amphibians not found elsewhere on the Outer Banks. Nags Head Woods offers prime habitat for herps—extensive acreage, sheltered areas protected by forested dunes, freshwater ponds, and a wide stretch of island with surrounding scrub vegetation. Much of the land is now within The Nature Conservancy's Nags Head Woods Ecological Preserve. In a 1987 survey, Alvin Braswell, State Museum curator in herpetology, found a total of 46 species of reptiles and amphibians here, about double the number of species found anywhere else on the Outer Banks.

The eastern box turtle, *Terrapene carolina*, is North Carolina's state reptile. Found throughout the state, it can live to be 100 years old. Its eggs are food for raccoons, foxes, and snakes. Courtesy of Alvin Braswell.

Alligator hunters in eastern North Carolina, ca. 1910 (*top*); H. H. Brimley with alligator (*bottom*); alligator skull with bullet hole (*opposite*). The American alligator, *Alligator mississippiensis*, is making a comeback after near extinction in the 1960s. The large crocodilian reaches the northern limit of its range at Albemarle Sound. Hunters and Brimley, North Carolina State Museum Archives; alligator skull courtesy of Rosamond Purcell.

Some of the species on the island represent disjunct populations, which were prevented by water and uncrossable terrain from making contact with related populations on the mainland. How did these animals come to be at Nags Head Woods? A few are probably descended from populations that were separated from the mainland by the forces that created the barrier islands.

In glacial times, sea level is lower than in warmer times, and the coastline of North Carolina extends further out on the continental shelf. When the climate warms, glaciers melt and water level rises. The last time this warming cycle began (around 10,000 years ago), rising water created the barrier islands, and some populations of animals were cut off from the mainland. The stable habitats of mainland-like forest and freshwater ponds at Nags Head Woods have probably sheltered the descendants of these populations ever since. Chicken turtles, oak toads, and pine woods snakes probably were among the original barrier island residents.

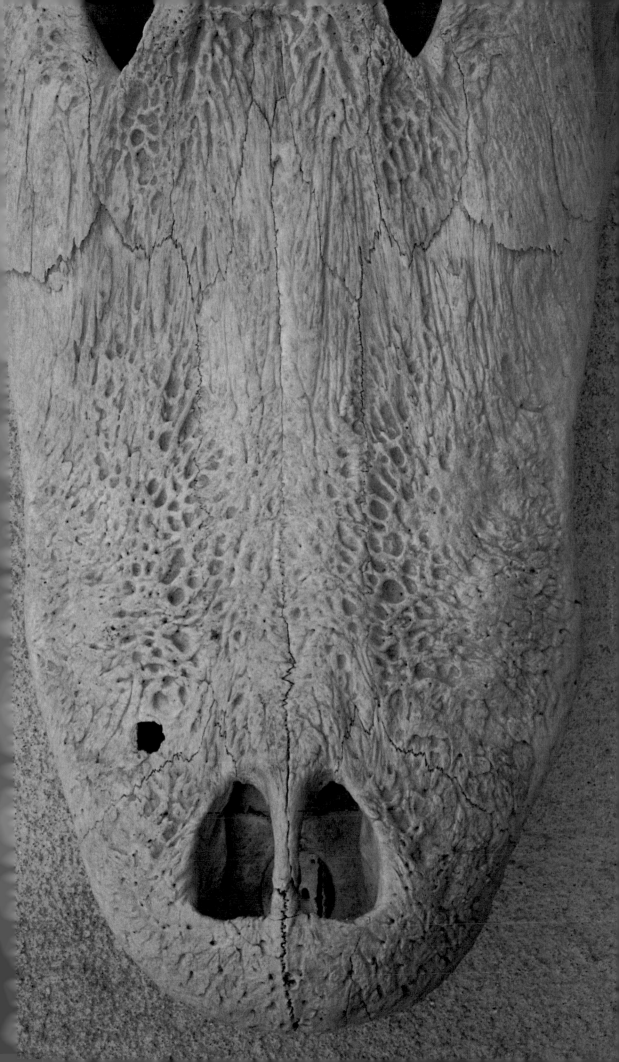

Eastern chicken turtle, *Deirochelys reticularia reticularia* (*above*); timber rattlesnake, *Crotalus horridus* (*facing page*). Nags Head Woods on the Outer Banks shelters many reptile and amphibian populations cut off from the mainland: green anole, Cope's gray treefrog, redback and marbled salamanders, eastern chicken turtle, timber rattlesnake, eastern worm snake, pine woods snake, and Carolina swamp snake. Courtesy of Alvin Braswell.

But most denizens of Nags Head Woods arrived in the millenia since the barrier islands formed, on rafts of logs and debris during periods of high winds or floods, or on land bridges that sometimes connect the island with the mainland. People were responsible for some reptile populations. Nags Head is one of the oldest resorts on the East Coast, attracting vacationers in increasing numbers since the Civil War. Tourists presumably aided the distribution of green and squirrel treefrogs, box turtles, green anoles, and yellowbelly sliders.

Though most of Nags Head Woods is protected in The Nature Conservancy's ecological preserve, rapid development alters habitat around the preserve, affecting water resources in the entire area. If development continues at the current pace, some herp populations will be reduced, and perhaps extinguished. The timber rattlesnake, which requires large forested areas to escape human interference, would likely be the first to disappear.

Because many herps are finely tuned to their surroundings, they can be especially vulnerable to habitat loss. Nowhere is this truer than in the vanishing longleaf pine woods of eastern North Carolina. Home to a great diversity of reptiles and amphibians, these longleaf pine communities include scarlet king snakes, tiger salamanders, pine snakes, pine woods snakes, and mimic glass lizards.

The Carolina gopher frog lives here. Like many frogs, it requires an ephemeral pond in which to breed—a shallow depression that fills with winter and spring rains, then dries up in the summer. Such a pool does not contain fish populations that would prey on tadpoles. Unfortunately these pools are not as common as one might expect. In 20 years of fieldwork on the gopher frog, Braswell found only a dozen active breeding sites remaining in North Carolina. Historically, the periodic fires of the pine woods helped to keep the naturally low places clear of shrubby vegetation, but when people suppress fires in this region, shrubs and small trees clog up the potential pools. Off-road vehicles can ruin ponds as well. Without a suitable pond, breeding stops. Within a few years of abstinence, the frog population declines and disappears.

Gopher frogs require even more specific breeding conditions than do other frogs. The best gopher frog sites are situated on large tracts of land dotted with multiple ephemeral ponds: Croatan National Forest, Camp Lejeune, the Green Swamp. Roads and cleared fields make travel perilous for frogs — so the fewer, the better. Multiple ponds in an area increase a frog's chances for breeding, because a particular pond may not be suitable every year.

South of North Carolina, off-season gopher frogs live in burrows made by the gopher tortoise, a turtle not found in this state. No one knows exactly where gopher frogs live in North Carolina outside of the breeding season. But during their brief breeding season in late winter and early spring, the male gopher frog reveals its location with an unnerving mating call that sounds like a deep, human-like snore. It is not the only call around; in the frog-rich Coastal Plain, 8 to 10 species may be calling at once. Volume counts in successful calling; females seem to prefer louder calls, perhaps indicating a brawnier mate. Discounting the percussive rubbings of insects, frog calls may well have been the first true vocalizations heard on Earth.

Once the gopher frogs find each other, the male fertilizes the female's eggs as she attaches them to a single vertical stem at the deep end of the pool. Developing from egg to juvenile frog takes two to four months, allowing

metamorphosis to occur before the pond dries up. The skin of the gopher frog is toxic, providing some protection from predators. Predators like raccoons sometimes get around this survival strategy with a tactic of their own. They skin the frog and dine on the non-toxic innards. Gopher frogs that make it through metamorphosis are relatively long-lived, and may be four years old before they breed.

Braswell found that one of the biggest threats to gopher frog survival is the existence of manmade barriers between breeding sites and the terrestrial habitat of adults. If long-term drought or lowering of the water table dries up breeding ponds for a few seasons, a gopher frog population is at serious risk if roads, clearings, and distance from appropriate habitat prevent it from seeking other breeding sites. Without a safe passage to and from neighboring sites, gopher frogs can't recolonize areas where local extinctions occur, and the range of the species may be reduced.

Many North Carolina amphibian populations suffer from loss of suitable habitat, but some species are losing numbers even though their territory remains fairly stable. The river frog seems to have disappeared mysteriously from the state. The mountain chorus frog was last recorded in 1946, and green salamander and southern dusky salamander populations are down.

The smallest and rarest turtle in North America has strict habitat requirements provided by few other states than North Carolina. The bog turtle makes its home in a cool, spring-fed bog. Fewer than 1,700 bog turtles live in the South, and most of those are found in the Mountain and upper Piedmont regions of North Carolina.

Bog turtle habitat may be lush with damp sphagnum mosses, needle rushes, and native cranberries. Cool black muck shields the bog turtle from predators and provides it with a banquet of salamanders, caterpillars, slugs,

Gopher frog, *Rana capito*. Courtesy of Alvin Braswell.

Bog turtle, *Clemmys muhlenbergi*. The bog turtle is nearly the smallest turtle in the world; at three to four inches long, it weighs less than a baseball. Courtesy of Jeff Beane.

earthworms, and crayfish. A map of bog turtle habitats plotted by Dennis Herman, State Museum curator of living collections, shows that North Carolina has twice the number of reported sites as other southern states.

Bog turtles live in temporary quarters. It is only a matter of time before a natural invasion of red maples, tag alders, and swamp roses transforms the bog into woodland, forcing the bog turtle to move on or perish. Biologists think that in the past, grazing herds of elk and buffalo cleared the bog habitat as they passed through mountain wetlands. Dairy cows performed the same role, and bog turtles continued to thrive. Now mountain valleys that harbor bogs are coveted for commercial development and polluted by agricultural runoff, and the bog turtle has been federally listed as a threatened species.

The bog turtle's tenuous hold on suitable habitat spurred Herman and the North Carolina Herpetological Society to help populations survive by preserving fragile mountain bogs. Project Bog Turtle works with landowners to save bog turtle sites from commercial development and to allow selective tree cutting to forestall woodland succession.

Reptiles and amphibians inspire dedicated study and even affection. The two herpetologists who initiated and organized the state's collection—C. S. Brimley and, later, William Palmer—contributed a combined total of 110 years to herpetological service. C. S. Brimley liked the small herps as much as his brother H. H. liked the big ones, and he compiled the first checklist of the reptiles and amphibians of North Carolina. He named the Neuse River waterdog (a.k.a. Carolina mudpuppy), an unusual salamander endemic to the Neuse and Tar River systems. A salamander that never leaves water, the Neuse River waterdog keeps its bushy gills into adulthood, and has three options for obtaining oxygen: with gills, through skin, or by gulping air at the water's surface with its lungs. Brimley's encouragement of young herpetolo-

gists was honored by teenaged collectors at the Washington Field Museum, who named a Coastal Plain frog they discovered *Pseudacris brimleyi*, Brimley's chorus frog.

A youthful penchant for snake collecting started many herpetologists on their careers. William Palmer, the State Museum's first curator of lower vertebrates, began in grade school to assist staff member Owen Woods in his care of the live snakes at the State Museum. As a teenager in the 1940s, Palmer accompanied museum director Harry Davis on collecting trips in eastern North Carolina. On one expedition Davis leaped from the car to collect a copperhead on the road. The snake struck Davis on the leg. To young Palmer's open amazement, Davis ignored the bite and calmly captured the copperhead. Only later did Davis reveal that his artificial leg came in handy during snake collecting. Palmer went on to catalogue and enlarge the collection, name a new North American reptile species, the mimic glass lizard, and coauthor with Alvin Braswell the definitive work on the state's reptile species, *Reptiles of North Carolina*.

People in eastern North Carolina often refer to glass lizards as "joint snakes." The name comes from a centuries-old belief that the broken pieces of glass lizard tail will later reunite. In 1709, John Lawson described a "Brimstone-Snake" from the area; Palmer believes it was a glass lizard. "They might as well have call'd it a Glass-Snake," wrote Lawson, "for it is as brittle as a Tobacco-Pipe, so that if you give it the least Touch of a small Twigg it immediately breaks into several Pieces. Some affirm that if you let it remain where you broke it, it will come together again."

Something about snakes has made them a part of local tradition throughout North Carolina. The beautiful, venomous copperhead snake bears the nickname highland moccasin, pilot, white oak, red oak, poplar leaf, chunk head, and adder, depending upon where in North Carolina it is found. Perhaps because of their longevity, live reptiles at the State Museum won the affection of visitors, and often received nicknames.

"Pamlico Pink was one of the most beautiful snakes I ever saw," wrote Dr. William Mann of the National Zoo. A canebrake rattlesnake from Pamlico County, Pamlico Pink was on display from 1935 to 1946. His portrait was painted for a snake identification guide issued to World War II GIs training in North Carolina. Onslow, a diamondback rattlesnake, was on live display for 15 years, and Columbia the cottonmouth fascinated visitors for many years. In 1964 a record 49-pound snapping turtle was captured near South Creek, christened Big Bad John, and carted to the State Museum for a popular living exhibit.

Live reptiles with nicknames brought in crowds, and gave visitors a way to connect with the animals. No animal in the state was better known than George the Python, a beloved 15-foot Burmese python that enlarged the State Museum's popular image as "the snake museum." George received scores of valentines and thousands of letters from visiting school children during his 25 years at the State Museum; 7,000 people stood in the rain for his "Coming

The venomous copperhead snake (*Agkistrodon contortrix*) is found throughout North Carolina, usually in or near wooded areas. Not an aggressive snake around humans, it prefers a diet of insects, amphibians, hatchling turtles, and small birds and mammals. Courtesy of Alvin Braswell.

(*top*)
A diamondback rattlesnake named Onslow lived at the State Museum in the 1930s. North Carolina State Museum Archives.

(*middle*)
Hired as a janitor at the State Museum in 1937, Owen Woods soon became known as the "snake doctor." He tended the museum's live reptiles for 20 years. North Carolina State Museum Archives.

(*bottom*)
George, a 15-foot Burmese python from Vietnam, is moved to new quarters by curator William Palmer, Commissioner of Agriculture Jim Graham, and CSM Dewey Simpson in 1978. North Carolina State Museum Archives.

Out Party." George arrived (already named) from Vietnam in 1964 with a Special Forces sergeant returning to Fort Bragg, North Carolina. Throughout the Vietnam War and for years afterward, military personnel and Vietnamese refugees came to visit the exotic snake. George's popularity was probably due partly to his great size, and partly because he symbolized North Carolina's unique cultural ties to Vietnam.

The practice of naming animals in the living collections ended in the 1970s. The shift to anonymity reflected a national trend in museum exhibits, away from featuring individual specimens toward creating habitat dioramas with many associated species. Exhibit designers took their lead from biologists, who study animals not as individuals but as populations interacting with each other and their environment. Museum guides refrain from projecting human traits on the live animals they share with visitors, trying instead to portray a more realistic picture of animal behavior in the wild. The inevitable question visitors ask? "What's its name?"

SEVEN   Birds

No eastern state has a greater diversity of birds than North Carolina. The state's location midway between north and south on the continental shoreline allows birds of the South to share habitats with birds of the North. The Blue Ridge Mountains are as far south as the tiny saw-whet owl ventures to nest, while the coastal swamps of North Carolina mark the northernmost range for the subtropical anhinga. Half of the currently recorded 420 bird species breed in the state. Around 75 species that nest elsewhere spend their nonbreeding season here, and others pass through on annual migrations. North Carolina claims 19 of the 27 orders of birds in the world.

Major migration flyways follow aerial trails along the Appalachians to the west and the Atlantic coastline to the east. In spring, woodlands of the mountains and Piedmont ring with calls of northbound songbirds. Birds may fly in from the West Indies and beyond, staying only long enough to nest. In the fall, barrier islands shelter huge numbers of land birds on their way to the Antilles and South America. The long line of estuaries running from the sounds of northeastern North Carolina to the Chesapeake Bay is a major East Coast wintering ground for ducks, geese, and other wildfowl.

The appeal of birds strikes deep for those who thrill to the sight of migrating flocks or the sound of a loon's call over the water. Birds bring out the taxonomist in people—what species is this, what group does it belong to, is it rare in these parts? Of all the animals, birds may be the most widely documented. Many people watch birds, record bird sightings, and share their re-

"Although striking in appearance, it would hardly have won its place in poetry and legend but for its cry, which is one of the wildest notes heard on our sounds or about wilderness lakes. Loud and far-reaching, it comes ringing across the water with a quality of unspeakable sadness" (Pearson et al., *Birds of North Carolina*, 1942).

Common loon, *Gavia immer*. Courtesy of Rosamond Purcell.

Saw-whet owl, *Aegolius acadicus* (*right, top*); anhinga, *Anhinga anhinga* (*right, bottom*); ruby-throated hummingbird, *Archilochus colubris* (*facing page*). North Carolina is at the northern edge of the breeding range of the anhinga, and the southern edge for the saw-whet owl. Many other birds, like the ruby-throated hummingbird, nest in the state in season and winter elsewhere. Courtesy of David Maslowski, David Lee, and Paris Trail.

ports with others. Verified records, along with data-rich specimens in the State Museum bird collection (skins, spread wings, nests and eggs, recordings, photographs, and videos) show changes in bird diversity and distribution over time.

Records for some state birds date back for centuries, adding invaluable information on species now threatened, endangered, or extinct. The long rec-

ord also reveals changes in our culture's complex relationship with birds, beginning with the delightful accounts of colonial naturalists.

Captain Arthur Barlowe documented the "cranes" of Roanoke Island in 1584. Two years later Thomas Hariot found on Roanoke Island, "turkey-cocks and turkey-hens, stock doves, partridges, cranes and herons, and in winter great store of swan and geese . . . also parrots [Carolina parakeet], falcons, and merlin-baws, which although with us they be not used for meat, yet for other causes I thought good to mention."

Collector John Lawson listed ten pages of bird species in his 1709 account, *A New Voyage to Carolina*, offering a glimpse of the abundant bird life of the Piedmont and Coastal Plain: "We killed of Wild Fowl, four Swans, ten Geese, twenty-nine Cranes, ten Turkies, forty Ducks and Mallards, three dozen of Parrakeeto's, and six dozen of other small Fowls, as Curlues and Plover, &c." Astonished by the now-extinct passenger pigeon, he wrote, "I saw such prodigious Flocks of these Pigeons in January and February, 1701–2 . . . that they had broke down the Limbs of a great many large Trees all over those Woods, whereon they chanced to sit and roost."

Carolina parakeets in the act of destroying orchard apples inspired William Byrd, surveyor for the Virginia North Carolina border in the early 1700s, to write, "They are very Beautiful; but like some other pretty Creatures, are apt to be loud and mischievous." William Bartram recorded in his

Ivory-billed woodpecker, (*top*), and passenger pigeon (*bottom*), paintings by ornithologist Alexander Wilson, the "father of American ornithology." Wison recorded North Carolina's sole documentation of the ivory-billed woodpecker in 1811. The extinct passenger pigeon was abundant in the state until the late nineteenth century. Courtesy of Janet Havens and the American Studies Program, University of Virginia.

Travels of the 1770s that in North Carolina the parakeet is "very numerous, and we abound with all the fruits which they delight in." In 1841, John James Audubon wrote of their beauty and abundance: "They present to the eye the same effect as if a brilliantly colored carpet had been thrown over them. . . . The gun is kept at work: eight or ten, or even twenty, are killed at every discharge."

The ivory-billed woodpecker once lived as far north as North Carolina, though its numbers may have dwindled by the 1770s when Mark Catesby recorded that the bill was worth two or three buckskins in trade to Canadian Indians, who fashioned the bills into coronets for nobles. John Lawson mentioned the bird's "tuft of beautiful scarlet feathers." No one left as definitive a portrait as did Alexander Wilson, a Scotsman known as the father of American ornithology. On his travels through the South in 1811 he captured an ivory-billed woodpecker 12 miles north of Wilmington.

> This bird was only wounded slightly in the wing, and on being caught, uttered a loudly reiterated and most piteous note, exactly resembling the violent crying of a young child; which terrified my horse so as nearly to have cost me my life. It was distressing to hear it. I carried it with me in the chair, under cover, to Wilmington. . . . I took him upstairs [in the Wilmington hotel] and locked him up in my room, while I went to see my horse taken care of. In less than an hour I returned, and, on opening the door, he set up the same distressing shout, which now appeared to proceed from grief that he had been discovered in his attempts to escape. He had mounted along the side of the window, nearly as high as the ceiling, a little below which he had begun to break through. The bed was covered with large pieces of plaster; the lath was exposed for at least 15 inches square, and a hole, large enough to admit the fist, opened to the weather boards; so that, in less than another hour he would certainly have succeeded in making his way through.

Wilson tied the bird's leg to a table with a string, and left to find food for it.

> As I reascended the stairs, I heard him again hard at work and on entering had the mortification to perceive that he had almost entirely ruined the mahogany table to which he was fastened, and on which he had wreaked his whole vengeance. While engaged in taking the drawing, he cut me severely in several places, and, on the whole, displayed such a noble and unconquerable spirit that I was frequently tempted to restore him to his native wilderness. He lived with me nearly three days, but refused all sustenance, and I witnessed his death with regret.

This feisty bird proved to be the first and only documented record of ivory-billed woodpeckers in North Carolina, and it established the state as the northern limit of this largest of American woodpeckers. Unconfirmed reports of sightings in eastern North Carolina continued into the 1970s.

Few recorded bird observations survive from antebellum North Carolina. Episcopal minister the Reverend Moses Ashley Curtis recorded and collected many birds in the state, but his "Birds of North Carolina" manuscript was never finished. After the Civil War, natural history became more and more the domain of trained specialists, whose extensive collections and publications defined their disciplines. Ornithologist and army doctor Elliott Coues published the *Key to North American Birds* in 1872, soon after a tour of duty at Fort Macon on the coast of North Carolina. Along with accounts of the state's warblers and shorebirds, Coues recorded the presence of ivory-billed woodpeckers on the mainland, but produced no specimen as documentation.

The curator of the bird collection at Harvard's Museum of Comparative Zoology, William Brewster, journeyed to the mountains surrounding Asheville in 1885. His trip journal documents some of the first records for the area:

> After a visit to the beautiful Cullasaja Falls, I spent an hour or so collecting in the woods nearby. At the Falls I heard the first *Contopus borealis* [olive-sided flycatcher]. In the impenetrable rhododendron bordering the river, Canada Flycatchers, Wilson's Thrushes, Wood Thrushes, and Black-throated Blue Warblers were singing. In the open hardwood timber above I found *Dend. blackburnae* [Blackburnian warbler], *D. pennsylv.* [chestnut-sided warbler], *Sphyrapicus varius* [yellow-bellied sapsucker] (a pair breeding), *Vireo Solitarious* [solitary vireo], *Hydemeles ludoviciana* [rose-breasted grosbeak], *Contopus virens* [wood pewee], etc.

Later, Brewster heard of John Cairns, a gifted young naturalist in nearby Weaverville who published one of the first checklists of birds in the area. Brewster lost no time in enlisting Cairns's help in collecting bird specimens for the Harvard museum. Cairns's study of the breeding habits of an Appalachian endemic subspecies, the black-throated blue warbler, was published after the young naturalist's accidental death on Mount Mitchell. The Cairns's black-throated blue warbler was named in honor of the hard-working Weaverville naturalist.

When State Museum curator H. H. Brimley heard of the untimely death of John Cairns, he secured funds from the State Board of Agriculture to purchase Cairns's entire bird collection for the State Museum. Brimley and Commissioner of Agriculture T. K. Bruner journeyed to Asheville to meet with Cairns's widow, but they were too late. She had permitted "a Boston scientist to pick over the entire lot & secure most of the very best specimens." Brimley and Bruner acquired around 50 of Cairns's specimens, but Harvard's Brewster took the bulk of the valuable North Carolina collection back home with him.

As self-taught naturalists, H. H. Brimley and his brother, C. S., published scores of articles on Raleigh bird life while still in their twenties, and contributed some of the first specimens to the State Museum's bird collection.

Museum curator H. H. Brimley and Commissioner of Agriculture T. K. Bruner photographed mountain scenes in southwestern North Carolina for use in tourism promotion. In Weaverville they attempted to acquire a valuable bird collection for the State Museum, but were beaten to it by the bird curator from Harvard's Museum of Comparative Zoology. North Carolina State Museum Archives.

H. H. Brimley's early passion was waterbirds; he collected the state's second record of a sooty shearwater in the summer of 1890. A hunter and an athlete, Brimley shot the shearwater as it flew over the surf near Fort Macon, then swam offshore to bring it in.

The shearwater may have ended up in one of the waterfowl displays that Brimley created for the North Carolina Department of Agriculture. Eager to promote the economic uses of its natural resources, the state included birds among the building stone and tobacco twists on display at expositions and trade fairs. In the late nineteenth century, tremendous flocks of waterfowl wintering in the sounds and waterways of eastern North Carolina supplied the meat markets of the Northeast. Among Brimley's exposition artifacts was an enormous pre–Civil War market gun, weighing 94 pounds with an eight-foot barrel, that could kill more than 50 birds at a time. The gunners decimated the American golden plover, recorded as common and "much sought for the table in the northern markets" by Curtis in the 1860s. Canvasback, redhead, teal, ruddy ducks, and bufflehead were among the ducks that hunters packed in barrels in remote waterways of North Carolina and shipped via steamer to Norfolk and points north for the tables of fine urban hotels. Wealthy recreational hunters also were beginning to discover the area, and the state saw value in promoting itself as a duck hunter's paradise.

To secure specimens for the waterfowl exhibit in the 1884 North Carolina Centennial Exposition, H. H. Brimley traveled by train, steamer, and oxcart to remote Coinjock, in the heart of the bird market-hunting industry off Currituck Sound. He bought birds from commercial hunters at the going rate: a pair of canvasbacks for a dollar, small ducks at four for a quarter. Brimley showcased aquatic birds "such as are of commercial value" at the 1893 World's Columbian Exposition in Chicago, and sent a loon, an anhinga, and quail along with his standard waterfowl case to the Paris Exposition of 1900.

When not on the road, these mounts became the first bird exhibits in the

State Museum. Some specimens remained on display for a hundred years: an anhinga and wood ibis from freshwater habitat, common and red-throated loons from inshore waters, a northern gannet from off the coast, and the first magnificent frigatebird taken in the state, collected in 1899 in Pamlico Sound. By 1907, Brimley wrote that his ark-like display included "among natural surroundings a collection of the different species of wildfowl found in the State. This case had a beautifully painted background and contained a specimen of the male and female of practically all of our native wild geese and ducks." Local illustrator and former curator Frank Greene painted exhibit backdrops for Brimley, who often shot his own specimens on hunts in eastern North Carolina.

Naturalistic habitat dioramas backed by painted landscape murals came into vogue in natural history museums here and in Europe around 1900. The twentieth-century natural history museum attempted to immerse the visitor in a sensual experience of nature that came close to "being there." Elaborate environments at the American Museum of Natural History and the National Museum of Natural History (Smithsonian) made the reputations of curators Carl Akeley, William Hornaday, and others, who hired well-known bird illustrators like Bruce Horsfall to paint their backdrops. Brimley's waterbird exhibit was modest compared with the habitat dioramas of the big-league museums, but his intentions were the same: to re-create for the urban museum visitor the sense of nature that he had experienced directly, spending long hours observing the habits of birds in the marshes and sounds of eastern North Carolina. Exhibits interpreted Brimley's conviction: "the sentimental and esthetic [nature of birds] has at least equal weight with that relating to dollars and cents. It is the living, pulsing bird that excites my interest."

Many of the prominent American curators of the time, including Akeley and Hornaday, were avid hunters. The sporting tradition — communion with nature, the robust outdoor life, the zest for manly adventure—was the message underlying museum exhibitry of the era. Nature offered a healthy antidote to the tension of city life, as in this poem by H. H. Brimley printed in the *Charlotte Observer* in 1916.

The wildfowl call from the marsh,
    The geese honk high in the air,
And here like a prisoner chained I sit
    Crouched down in my office chair.
And O! it's so quick the trail I'd hit
    Were it not for this business care.

The dreary day has gone by,
    My door lock snaps like a bell;
I'm off to the marsh ere daylight comes—
    And my work may go to Hell!
The Grip of the Wild is round me fast—
    I'm free from prison cell!

Brimley often escaped to legendary Camp Bryan on Lake Ellis, a hunting club favored by Babe Ruth, Lou Gehrig, and Ted Williams. Hunting clubs became a thriving business along the sounds and barrier islands, attracting the rich and famous to prime duck and geese wintering sites and providing income for local people as guides and decoy carvers.

Brimley's connections with prominent sportsmen in the state did produce some rare specimens for the State Museum's collection at a time when hunting threatened some of the state's most beautiful birds. The Currituck Shooting Club donated "several fine and rare species of water-fowl" to the museum in 1896. Some curators of the time defended the shooting of a rare animal if the specimen ended up in a museum. Collections preserved for posterity animals that seemed to be headed for inevitable extinction.

Brimley had access to the private bird sanctuary at Orton Plantation near Wilmington, where conservationist T. Gilbert Pearson collected the first state record of an anhinga in 1898. Here Brimley collected great egrets and snowy egrets for display. He wrote, "Those birds of the beautiful plumes that have caused their downfall almost to the point of extinction . . . are shown with nests, eggs, and young, all the specimens coming from the only place in the State where they still exist, on the strictly protected lands and water of Mr. James Sprunt, of Wilmington, who most kindly allowed the collection and permanent preservation of these most beautiful creatures while there was yet time."

The turn-of-the-century trade in "aigrettes" for ladies' feathered hats took a devastating toll on the coastal birds of North Carolina. Feathers of gulls, terns, and egrets were prized, and pelicans, eagles, and vultures were sometimes sought for hat plumes. Elliott Coues reported the least tern as the most

abundant tern of all at Fort Macon in the 1870s, but by Brimley's day the least tern was almost nonexistent in the area. "Thither went the plume hunters, and season after season butchered these exquisite creatures to get the wings for the New York millinery trade," wrote Pearson. Hunters took 10,000 skins from one tern nesting site in a single season. The beautiful wood duck suffered the same treatment.

A sense of urgency about the fate of birds gave birth to the conservation movement, as some citizens began to wonder if the vast flocks of waterfowl and shorebirds could go the way of the panther, beaver, and wolf. The issue of wildlife conservation, particularly for birds, led Pearson, Brimley, James Y. Joyner, and other concerned North Carolinians to found the Audubon Society of North Carolina for the Study and Protection of Birds. Pearson later went to New York to lead the National Association of Audubon Societies.

Bird conservation was a volatile issue, affecting the variety of ways that birds figured into the culture and economy of the state. H. H. Brimley wrote, "The sportsman looks at things from a sportsman's standpoint entirely but he is not the only pebble by any means. The Currituck market gunner has his point of view and must be considered and the poor man who shoots robins as game is at least entitled to a hearing." The hunting of migratory species was an honored custom in many North Carolina communities. On the coast, residents hunted loons each spring near Cape Lookout, and a meal of "Harkers Island turkey," complete with onions and cornmeal dumplings, was as traditional as Thanksgiving.

Knowing the cultural resistance it faced, the Audubon Society attempted to draft passable legislation, dubbed "the bird bill," which would protect nongame and game birds. Governor Aycock leant his support to the bill, which the North Carolina General Assembly approved in 1903. The killing or

harming of any species not specifically exempt from protection by the bird bill was outlawed, and a state game commission was formed to regulate hunting of game birds. While the power of the bird bill was chipped away by subsequent legislation, this early conservation effort paved the way for successful wildlife management after World War I.

As interest in birds and nature study grew, people from across the state regularly sent specimens, nests, eggs, and records of sightings to the State Museum. From the growing collection, Pearson and the Brimley brothers compiled records, wrote bird descriptions, and commissioned illustrations for *Birds of North Carolina*, one of the few state bird books when it was published in 1919.

The 1942 edition of *Birds of North Carolina* noted strides in bird conservation in the twentieth century, but chided suburbanites for their treatment of

Canada warbler and American redstart (*right*); kinglets, nuthatches, chickadees (*above*). The 1919 edition of *Birds of North Carolina*, one of the first bird books in the South, included art by Robert Bruce Horsfall (warbler and redstart). Roger Tory Peterson (kinglets, nuthatches, chickadees) donated his paintings and drawings for the 1942 edition.

the "Nighthawk or Bull-bat . . . which congregated in large numbers on summer evenings, in suburban districts, to get their evening meal of insects. This bird was a favorite target of misguided sportsmen." At dusk, flocks of chimney swifts flew in to roost in the vents and chimneys of many downtown and suburban buildings. H. H. Brimley recorded the odd daytime occurrence of "great numbers of swifts gathering and pouring down the chimney of the heating plant" of the museum building during a solar eclipse in the 1920s. The authors listed regional names for certain birds, some of which linger in vernacular speech today. The Florida cormorant was known as a Bogue lawyer; the pied-bill grebe as a didapper or hell-diver. The scorned wood ibis was called flint head, gourd head, and gannet. The anhinga was commonly known as water turkey and snake bird, and in some African-American communities, as the whang-doodle.

Since the 1930s, volunteers and ornithologists have recorded banding reports and observations in *The Chat*, the periodical of the Carolina Bird Club. The club issued *Birds of the Carolinas* in 1980. Volunteers helped to chart the success of comeback species like the bald eagle, brown pelican, and wild turkey, and mapped nesting sites for the *N.C. Breeding Bird Atlas*.

Documentation for birds in the bird collection of the State Museum shows just how dynamic the state's bird populations really are over time. Records indicate a rise in the total number of species breeding in North Carolina. At the turn of the twentieth century, 150 breeding bird species were reported in the state. Now the figure hovers around 210 species. Close to half of the breeding birds in the state expanded their ranges in the last hundred years. Each species has its own reasons for expanding its range, but the well-documented tree swallow demonstrates how one species swiftly exploited manmade changes in the landscape.

In 1979, Harry LeGrand and Eloise Potter, a veteran worker in the State Museum's bird collection, found North Carolina's first tree swallow nest in an abandoned woodpecker cavity along the New River in Ashe County. A second nest was discovered a year later on the French Broad River north of Asheville. Volunteers working on the N.C. Breeding Bird Atlas found more nests in the mountains of southwestern North Carolina. Nests began to turn up in the Piedmont in trees killed when river bottoms were flooded to create Lake Jordan. In the Coastal Plain, tree swallow nests appeared in bluebird boxes and in dead trees in flooded areas.

A hundred years ago, tree swallows were considered to be midwestern birds that passed through North Carolina during spring and fall migration seasons. Nesting tree swallows gradually moved into open, manmade spaces—agricultural clearings, reservoirs and flooded areas, developed lands—dotted with dead trees, or snags, that furnish the birds with ideal nesting cavities. Typically, nesting tree swallows look for a roomy wood-pecker hole in a dead tree, where small colonies of a dozen pairs of tree swallows may build their nests. The boom in North Carolina's beaver population in the late twentieth century most likely assisted the arrival of breeding tree swallows, because beaver dams flood lowlands, creating snags. In addition, many tree swallows are now imprinted on bluebird boxes as nesting sites. Tree swallows simply took advantage of a favorable situation—nesting sites furnished by various human activities—and joined the number of breeding birds in North Carolina.

An increase in diversity may actually push some "oldtimer" species into decline. Newly arrived species have put pressure on some endemic mountain birds. The southern Appalachians are home to a handful of endemic sub-species believed to be relics of the last ice age. In those cold times, spruce-fir forests covered the Southeast, which was populated by birds currently found farther north. When the climate began to warm up around 10,000 years ago, hardwoods displaced the southern spruce-fir forest, except in the cooler microclimates of southern Appalachian peaks. Bird populations became separated—most of the population continued living in northern spruce-fir forests, but pockets remained on southern peaks, where their de-

scendants live today. These isolated populations represent subspecies found nowhere else.

The Appalachian Bewick's wren is an endemic species of the Mountain region that may now be gone. When Professor Brewster visited the area in 1885, he reported the wren nesting in outbuildings all around Asheville. However, its numbers declined by the 1930s, and by the 1950s most of these wrens were nesting only in higher elevations away from towns. The last breeding pair was documented on the Blue Ridge Parkway near Mount Pisgah in 1971. Appalachian Bewick's wren is believed to be extinct now, possibly pushed to the brink by the house wren, a species that expanded its range in North Carolina in the twentieth century.

Development in the mountains created open habitats that favored new species of birds along utility rights-of-way, road cuts, and logged woods. Many species now regarded as abundant in the mountains—barn swallows, starlings, brown-headed cowbirds, house wrens—are relative newcomers, attracted by cleared land and second-growth forest. Meadowlarks, horned larks, and woodcocks moved into farm pastures; chestnut-sided warblers favor logged-out forest.

One endemic species, Cairns's black-throated blue warbler, has increased in numbers in spruce-fir forests that have been damaged by an insect called the balsam woolly adelgid, itself a newcomer introduced in the 1960s. However, loss of Fraser firs and old-growth forest has diminished the ranges of six other endemic birds. The Appalachian black-capped chickadee now exists in small, isolated populations at the highest peaks.

North Carolina's geographical location created a very different habitat at the far eastern edge of the state. An international congress of seagoing birds holds forth about 45 miles off the coast of Cape Hatteras. On an autumn day, birds that breed in Greenland, the South Polar islands, the Alaskan tundra, the Mediterranean Sea, islands off the coast of west Africa, the Caribbean, and the southeastern United States gather to feed in the same waters. Deep-sea fishermen know this stretch of water as The Point. Bird watchers claim it is one of the best places to see pelagic (open sea) birds in North America, and the state Audubon Society designated the area as one of the first Important Bird Area habitats in North Carolina.

Shaped by ancient geologic processes, the edge of the continental shelf is a crossroads where cold north Atlantic waters meet warm subtropical waters. Wildlife from both regions moves in and out during the year. The Point is the southernmost site where high concentrations of kittiwakes gather, and it is the only place where the highly endangered Bermuda petrel has been spotted at sea. The habitat provides rich offshore feeding grounds for at least 50 species of seabirds.

Little was known of the birds in this rare habitat until David Lee, State Museum curator of birds, began a study of species diversity here in 1975. Working from chartered fishing boats in every month of the year, Lee and his colleagues discovered the regular presence of more than 20 species that were previously undocumented or believed to be accidental in North Carolina waters. New questions naturally arose. What role does The Point play in the mysterious yearly cycle of these global birds? Are their populations vulnerable to disturbances in the area—oil pollution, drilling activity, heavy-metal wastes? To find out, Lee and colleagues collected data on feeding habits, molt patterns, age, weight, body temperature, and sex ratios. They examined structural variations, parasites, and mercury concentrations in the birds' tis-

Canada warbler, *Wilsonia canadensis*. Developed land opens new habitat in the Mountain region for barn swallows, starlings, house wrens, and other birds, which have crowded native woodland birds such as this Canada warbler. Courtesy of David Lee.

sues and feathers. In the process, the State Museum's seabird collection became a resource for international researchers studying this wide-ranging group of birds.

From the data a new model emerged of a heavy-traffic ecosystem that over the year hosts populations of seabirds from both southern and northern breeding ranges. An abundant species that breeds in South America, the greater shearwater numbers around five million breeding birds. Practically the entire global population of greater shearwaters passes through North Carolina waters in May and June. Wilson's storm petrel is another familiar sight to offshore fishermen; called Mother Cary's chickens, or just "chickens," Wilson's storm petrel may be the most plentiful wild bird in the world. Oldtime whalers picked up loads of "chickens" killed by collisions with ship rigging and used their oily carcasses for onboard candles.

Numbers of the uncommon Manx shearwater seem to be increasing at The Point. New breeding colonies in Newfoundland, founded by pioneering Manx shearwaters from Europe, may be the source of the growing population in North Carolina. In recent years the Atlantic puffin, a bird once decimated by hunters for the plumed hat trade, has been reported regularly by fishing boat captains at The Point. Puffin numbers may be rising because of National Audubon Society efforts to protect puffin breeding sites in Maine.

Tropical breeding birds generally raise far fewer young than do cold-climate species. All the Caribbean and West Indian breeding birds put together amount to less than 20 percent of the birds from one large colony of Leach's storm-petrels, a northern breeder. When tropical breeding sites are threatened by land-clearing, introduced predators, or hunting, these low-density breeders may become endangered. Total world populations of the Bermuda petrel, Madeiran petrel, and Trinidade petrel have been reduced to several hundred pairs or less. Their numbers at The Point may be few, but probably represent a significant portion of the global populations.

The black-capped petrel was long considered extinct, but nesting colonies on nearly inaccessible cliffs were discovered in Haiti in 1961. Now gone from four of the five islands where nesting sites previously occurred, black-capped petrels are down to 1,000 to 2,000 breeding pairs, a 50 percent drop in population since they were re-discovered. Lee found a high concentration of these birds at The Point during most of the year, and in May, August, and around New Years Day they are as common as cardinals at a bird feeder. The low breeding population and high numbers of adult males at The Point even during breeding season suggest that this winter-breeding bird must be commuting between Haiti and North Carolina to forage for food for its young.

The Bermuda petrel numbered in the hundreds of thousands in the sixteenth century, but was thought to be extinct by 1620, wiped out by hunting and introduced predators. In 1951, researchers discovered 18 pairs nesting on five tiny islets off Bermuda. Protected by the Bermudan government, the species is gradually increasing in number. One of the most endangered birds on

White-tailed tropic-bird, *Phaethon lepturus* (*above*), and Audubon's shearwater, *Puffinus iherminieri* (*left*). Birds from all corners of the Western hemisphere spend part of their year off the coast of North Carolina, where cold North Atlantic currents meet the warm subtropical Gulf Stream. Courtesy of David Lee.

Earth, the Bermuda petrel is known in only one place outside of Bermuda—The Point.

Throughout the year, species both rare and common move in and out of this offshore region. North Carolina coastal waters host the most diverse, and perhaps the most dynamic, seabird fauna in the western North Atlantic. Because a large percentage of the populations of many pelagic birds spend part of their year at The Point, these species are vulnerable to manmade disasters such as oil spills, and to long-term degradation of this valuable global habitat.

Our awareness of the wide-ranging birds of North Carolina expands with every byte of newly collected data. Brimley's nineteenth-century waterbird display had no black-capped petrels, no Manx shearwaters, no Atlantic puffins; Cairns's Appalachian bird checklist included the Appalachian Bewick's wren and passenger pigeon, now gone. Keeping track of the state's changing bird life requires the work of hundreds of people—careful collectors of essential information—whose devotion is constantly renewed by the fascinating nature of birds.

Mammals

Despite its remarkable success, Mammalia is the vertebrate class with the fewest species. But this class includes the largest animals living on Earth, and North Carolina has had its share of big mammals. Some, like the gray whale, buffalo, elk, and panther, have, because of hunting, disappeared from the state over the last 300 years. On the other end of the size scale, the pygmy shrew (⅛ ounce) and the eastern pipistrelle bat (⅕ ounce) still survive in the state. Mammals exhibit a range of extraordinary abilities: hibernation and migration; echolocation; the ability to stalk, leap, swim, gallop, even fly. The state's only marsupial, the opossum, uses its opposable big toes and prehensile tail to climb trees.

Mammals are warm-blooded animals with hair, milk-producing glands, and a well-developed cerebrum. Many are nocturnal, small, and/or well-camouflaged, making them hard to observe in nature. Most have teeth and jaws adapted for a specialized diet. Teeth may serve other purposes as well; for instance, beaver teeth are versatile construction tools. Characteristics of the skull and teeth help taxonomists classify mammals into orders, families, and species.

Naturalist C. S. Brimley authored the first systematic account of North Carolina's mammals in 1908. He acknowledged that mammals, especially the smaller rodents and insectivores, were harder to study than other vertebrates because of their nocturnal habits. Still Brimley managed: "From about 1888 to 1900 my brother [future State Museum director H. H. Brimley] and

"The Finner Whale, or Finback, is probably the most abundant of the larger whales, and it is the one most frequently found washed up on the beach . . . at points along the North Carolina coast" (H. H. Brimley, 1940).

Fin whale, *Balaenoptera physalus*, bone. Courtesy of Rosamond Purcell.

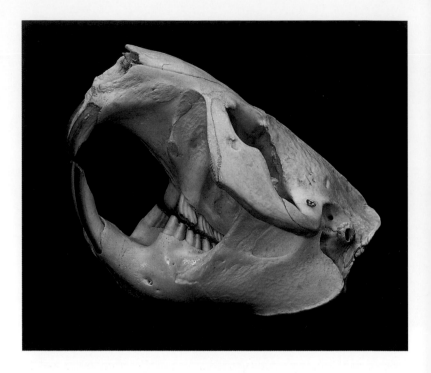

myself set out quite a number of traps and caught over a thousand mice and shrews, most of which were put up as museum skins."

In 1944, C. S. Brimley listed 65 land species and 15 marine species from the state. Now at 110 native mammal species, North Carolina has the most diverse mammal fauna east of the Mississippi River. Ten orders are present today, but in past times other orders occurred as well. In the Pleistocene epoch, giant ground sloths, woolly mammoths, and mastodons roamed the state. Mastodons and other large mammals were hunted by pre-Columbian cultures.

Around 30 species of marine mammals are permanent residents or seasonal migrants in North Carolina's waters. Southern and northern marine mammals coexist near the edge of their respective ranges off the coast, where warm tropical currents meet cold currents from the North Atlantic. Harbor seals and hooded seals wander south, and the Florida manatee visits southeastern North Carolina in the summer—pushing the northern edge of its range.

The first manatee documented from the state was captured in 1919 in Masonboro Sound near Wilmington. Called a "cowfish" by local watermen, the manatee created such a stir that the Howard and Wells Amusement Company bought the animal for public show. When cold weather arrived, the tropical animal sickened and died. A local businessman contacted the State Museum about the unusual mammal, and H. H. Brimley reported on the manatee's presence in the *Journal of Mammalogy*. The amusement entrepreneur donated the manatee's partially preserved hide to the State Museum, but Brimley could not salvage this first state specimen.

Manatees are now known to occur sporadically during warm months from the Cape Fear River near Southport to the Pamlico Sound off Ocracoke Island. The museum's mammal collection is part of a network of facilities that are repositories for manatee specimens, aiding research on this endangered mammal. One of four living species in the order Sirenia, the manatee is the only sirenian surviving in North America.

Two orders of cetaceans — the whales, dolphins, and porpoises — include 26 species living in North Carolina's waters. Apart from body size, the shape, size, and number of teeth distinguish marine mammal groups. One group of whales has baleen plates instead of teeth.

The first specimen to enter the State Museum's mammal collection was a baleen whale, acquired as a byproduct of a tradition now long gone on the Outer Banks. Cape Lookout was the last southern post on the East Coast for shore-based whaling. Until the early 1900s, families on the Banks hunted right whales, using distinctive long rowboats designed for carrying a crew of six out to sea. The Banker families counted on the return of migrating right whales, which swam close to shore by Bogue Banks and Shackleford Banks every spring. Families joined together to process the whales on the beach, sharing income from the sale of oil and whalebone.

Six boats pursued the big right whale that Absalom Guthrie sighted from the shore of Shackleford Banks on May 4, 1874. In a legendary battle, the whale struggled with the whalers for half a day in rough waters before it was killed. In keeping with the Banker tradition of naming a catch, the men christened the great whale "Mayflower," for the month in which it was taken. The payoff for the whalers was 40 barrels of oil and 700 pounds of whalebone.

Beaver, *Castor canadensis*, skull (*opposite*), and red-backed vole *Clethrionomys gapperi*, specimen series (*below*). Courtesy of Melissa McGaw and Rosamond Purcell.

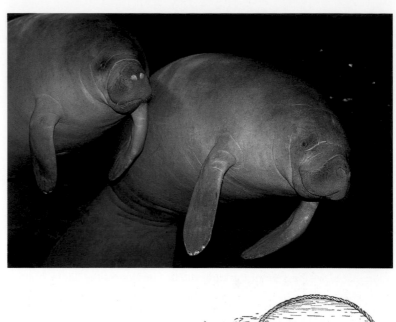

The skeleton was preserved by the Atlantic and North Carolina Railroad, which eventually donated the bones to the State Museum as an example of the natural resources of Carteret County. The bones lay on the floor under exhibit cases until 1894, when H. H. Brimley was hired to mount them. He scrubbed twenty years' accumulation of dirt, oil, and visitors' tobacco juice from the bones before he could assemble and mount the skeleton. Still on display in the State Museum, Mayflower's 50-foot-long skeleton has awed visitors for more than a hundred years.

The northern right whale has become the most endangered of all the great whales. Early whalers named it the "right" whale—the best whale to catch because it lives close to shore, swims slowly, and floats when dead. Right whale blubber was rendered into fuel oil. Baleen plates, the sieve-like plates that grow from the whale's upper jaw, were sold for buggy whips and corset stays. Though protected in the latter half of the twentieth century, populations of northern right whales are rapidly declining. The species' naturally low reproduction rate has not kept pace with the number of right whales

Beneath cases in the State Museum's Geological Room lay the bones of a right whale, named Mayflower by the whalers who captured it in 1875. H. H. Brimley articulated the skeleton for display in 1895. North Carolina State Museum Archives.

lost to illegal whaling, entanglement in fishing nets, and collisions with ships.

Sperm whales frequent deep waters off the continental shelf of North Carolina, and are seldom seen inshore. Storing oxygen in their tissues, sperm whales can stay underwater for up to an hour and dive more than a mile deep. In 1928, a beached sperm whale attracted thousands of tourists to the shores of Wrightsville Beach. When the great body began to decay, it was nicknamed "Trouble." With great difficulty H. H. Brimley managed to haul the sperm whale skeleton to Raleigh, where it was cleaned, mounted, and displayed at the State Museum. Generations of North Carolinians marveled at its size, and Trouble the sperm whale became the museum's official symbol.

Trouble and Mayflower now hang side-by-side on exhibit, along with the skeletons of a humpback whale, a True's beaked whale, and a blue whale. Only a handful of museums in the world exhibit as many whales, or maintain as diverse a collection of marine mammals as does the state of North Carolina.

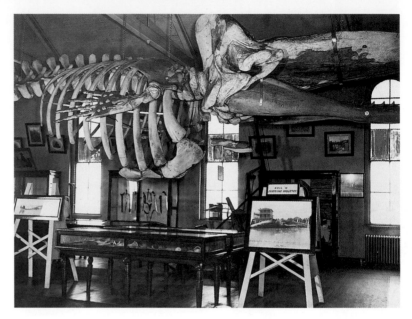

Beached sperm whale at Wrightsville Beach, 1928 (*above*). Nicknamed Trouble, the whale carcass was obtained by H. H. Brimley, who cleaned and mounted the skeleton in the State Museum (*right*). North Carolina State Museum Archives.

Chiroptera, the order of bats, is another focus of the museum's mammal collection. Found in every region of the state, all 16 species of North Carolina bats eat insects. A single bat may consume 21,000 insects a year—which translates as ¼ to ½ of its body weight daily. A colony of 150 big brown bats (North Carolina's common urban bat) may protect crops from 18 million corn rootworms every summer. When the flying insect population declines in autumn, most North Carolina bats hibernate, alone or in colonies, or they migrate.

Though one-third of North Carolina's bat species are on federal or state lists of endangered, threatened, or at-risk species, lack of field research leaves in question the true population status of some species, especially the reclusive forest bats. Until the 1980s, a scattering of records documented only isolated populations of Rafinesque's big-eared bats from the mountains and the Coastal Plain, usually near or in abandoned farm buildings. Recently Mary Kay Clark, the State Museum's curator of mammals, launched a survey of big-eared bats in southeastern North Carolina to determine where the bats occur in the state, what their roosting requirements are, and if the species is indeed rare.

Equipped with a quiver of research tools from bat-sized radio transmitters to tiny photochemical tags, Clark's research team found tree roosts of Rafinesque's big-eared bats in old-growth swamp forests along the Roanoke River and other southeastern bottomlands. Large cavities in ancient swamp tupelo, sycamore, sweet gum, and tulip poplar trees shelter the bats, which forage for insects in mature forests. The big-eared bat flies slowly, sometimes hovering like a hummingbird, and prefers to forage within the forest, not out in the open.

At one time, Rafinesque's big-eared bats probably ranged over the state, roosting in the cavities of huge old trees in mature forests. But mature forests have become rare, and appropriate tree cavities are scarce. Bats may find shelter and nursery space in manmade structures like buildings and bridges. Populations of forest-dwelling species like Rafinesque's big-eared bat and the southeastern bat may be on the decline as old-growth forests disappear. Although Clark's study discovered new big-eared bat colonies, the number of individuals in every monitored colony declined over a ten-year-period. Some colonies disappeared entirely. Rafinesque's big-eared bat is currently listed as a threatened species in North Carolina. Armed with a knowledge of the bat's roost requirements, land managers should be able to identify and protect areas key to the species' survival.

Several northern and southern bat species approach the limits of their ranges in the state. The southeastern bat, Seminole bat, Brazilian free-tailed bat, and one subspecies of Rafinesque's big-eared bat do not regularly occur further north than the Coastal Plain of North Carolina and southeastern Virginia. The rare small-footed bat and endangered Virginia big-eared bat are both at the southern edge of their ranges in North Carolina. Bat Cave in Henderson County harbors several species, including the critically endangered Indiana bat. Several Indiana bats were recorded in 1999 on national forest

The State Museum's Coastal North Carolina exhibit displays the skeletons of five whales, including the right whale Mayflower and the sperm whale Trouble. Courtesy of Peter Damroth.

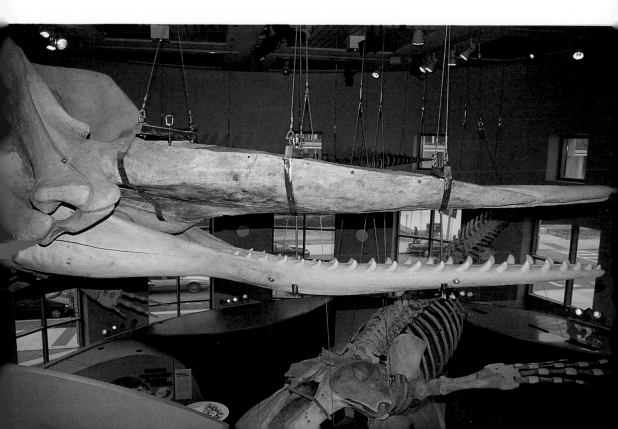

lands in Graham County, including the first record of an Indiana bat maternity colony from the southern end of the species' range.

The Mountain region harbors the greatest diversity of mammal species in the state, due in part to the wide variety of mountain habitats. In spruce-fir forests on Mount Mitchell, David Adams, chief curator at the State Museum in the early 1960s, found red-backed voles, smoky shrews, masked shrews, shorttail shrews, deer mice, and red squirrels. Adams also reported a southern bog lemming, a rodent usually found in moss-sedge bogs and other damp places in the region.

Some mountain mammals are descended from cold-climate animals that remained in the cool upland region when the Southeast warmed at the end of the last ice age. The region marks the southern boundary for some species of the order Insectivora, the shrews and moles. The eastern spotted skunk, at the eastern edge of its range in the mountains, prefers open forest habitat. Exterior markings distinguish the species from the striped skunk of the Piedmont and Coastal Plain, as does the gymnastic delivery of its powerful musk. When alarmed, the spotted skunk stands on its front legs with its tail arched

forward. In this handstand pose it can accurately direct the released spray, reported to be even more pungent than that of the striped skunk.

Until the mid-twentieth century in North Carolina, the woodchuck occurred mostly in the Mountain region. Locally known as a groundhog or whistle-pig, it was hunted for its fatty meat and its hide, which made a decent banjo head. In the last fifty years the species began to spread down from the hills into northern Piedmont counties. Highways and utility rights-of-way that serviced the region by the 1940s probably created the open habitat needed for the groundhog to extend its range. Actually a ground squirrel in the order Rodentia, the groundhog has now been recorded in the Coastal Plain. The cultivated fields and fencerows in every region are home to the burrowing groundhog, which grazes on grasses and garden produce.

Beavers, the state's largest members of the order Rodentia, range from

Spotted skunks, *Spilogale putorius*. The Mountain region has the greatest diversity of mammals in the state. The spotted skunk is at the eastern edge of its range here. Courtesy of Rosamond Purcell.

Mountain region to Piedmont to Coastal Plain. In his 1709 account, *A New Voyage to Carolina*, John Lawson wrote of an abundance of mischievous beavers who could be tamed, but were sometimes guilty of "blocking up your Doors in the Night, with the Sticks and Wood they bring thither." Heavy trapping eliminated the state's native beaver population by the end of the nineteenth century. According to C. S. Brimley, who collected the type specimen of the southern beaver, the last native beaver was reported from Stokes County in 1897. Forty years later, wildlife managers introduced Pennsylvania beavers in the Coastal Plain. More beavers were released in the Piedmont in following years. Regulated trapping allowed these populations to grow, and beavers now thrive along all of North Carolina's major river systems. Shallow beaver ponds provide a watery habitat for other animals, including muskrats, wood ducks, and kingfishers.

White-tailed deer were all but gone from the state by the early twentieth century, except in the Pisgah National Forest and coastal river valleys—the end result of all-season, no-limit hunting. Gamekeepers at the Biltmore Estate near Asheville brought in deer to stock the herd in the Vanderbilts' private game preserve, an action which eventually helped to boost mountain deer populations elsewhere. The North Carolina Wildlife Resources Commission began a deer restoration effort after World War II. Now extremely abundant throughout the state, white-tailed deer have tripled in numbers since 1980, topping one million in 1995.

Bears and panthers once roamed the whole state until settlers in the Piedmont eliminated that region's populations through overhunting and destruction of habitat. Eighteenth-century naturalist Mark Catesby described their original habitat in *Birds of Colonial America*: "[O]n the banks of these [up-land] rivers extend vast thickets of cane, of a much larger stature than those before mentioned, they being between twenty and thirty feet high, growing so close, that they are hardly penetrable but by bears, panthers, wild cats, and the like."

Reports of panther sightings still come to the State Museum from the mountains and Coastal Plain, but the last verified sighting of a panther in North Carolina occurred in the 1880s, when the panther's major food source, the white-tailed deer, declined. No specimens of North Carolina panthers exist in collections today. In eastern North America the only recognized population of panthers is found in southern Florida.

Around 9,000 black bears now live in the state's mountains and Coastal Plain—triple the number in the 1950s. The northeastern Coastal Plain region has become one of the nation's prime hunting grounds for black bear.

Much of the data on populations of game species, like bear and deer, and furbearers, like fox and beaver, has come from hunters and trappers. Skins of rabbit, otter, beaver, and mink entered the mammal collection through commercial trapping activities. Promoted by the early State Museum, tanning of native skins for leather became a commercial industry in rural areas of the state in the nineteenth century.

Hunters and others in North Carolina regularly send albino animals—mink, mole, raccoon, opossum, rabbit, and white-tailed deer—to the State Museum. Born without color pigment in their skin and eyes, albinos can be found in all animal groups. An exhibit of a few of these rare animals fascinated visitors to the State Museum for decades. In eastern North Carolina, some traditional healers use the madstone, or bezoar, a calcareous lump found in the stomach of a hoofed mammal. Believed by some to cure bites from venomous or rabid animals, a bezoar from an albino deer is considered highly potent and quite valuable. In Cherokee tradition, the chiefs of the Deer and the Bear tribes are white, and albino animals are considered sacred.

Native North Carolina beaver (*Castor canadensis*) specimen. Beavers were brought to North Carolina from Pennsylvania and other states to begin new populations after the last native beavers were taken by hunters in the 1800s. Courtesy of Jim Page.

The hides of mammals, stacked like cordwood at the Junaluska Leather Company, provided income for trappers and processors in the nineteenth century. North Carolina State Museum Archives.

(*opposite*) Albino deer, *Odocoileus virginianus*. Courtesy of Rosamond Purcell.

At one time in North Carolina, mountain meadows and upland Piedmont valleys were the grazing grounds for herds of American elk and buffalo, both extensively hunted in the 1700s. By 1765, William Bartram reported that "the buffaloe, once so very numerous, is not at this day to be seen in this part of the country." The last wolf was reported in 1887 in Haywood County.

At the turn of the twentieth century, the major natural history museums in the United States drew crowds with stunning dioramas of buffalo and moose, depicting the last vestiges of American wilderness. President Teddy Roosevelt, big game hunter and conservationist, captured the public imagination as the embodiment of the intrepid sportsman. Hunt clubs sprouted among the affluent classes, encouraging members to deliver trophies to museums for scientific study. The "Heads and Horns Museum" in New York copied a collection in the British Museum, and most museums acquired specimens from hunting trips.

Like many museum curators of the era, H. H. Brimley thrived on communion with nature, specimen collecting, and the thrill of the sporting life. Brimley wrote, "I have hunted deer in eastern North Carolina every season for more than twenty-five years . . . with some degree of success. . . . The hunting of this elusive animal carries with it thrills that are not found in the following of any other game bird or animal." Ecologist Eugene Odum described Brimley's approach to hunting: "Even the unsuccessful [hunting] trip is a success to the true sportsman if he be also a nature lover. Such a man was H. H. Brimley."

Brimley was determined to bring to the State Museum all of the lost land mammals of North Carolina. From the West he acquired an elk and a gray wolf. He bought a bull buffalo for $400 in 1900 from the Flathead Indian Reservation, along with a Rocky Mountain lion ($12.50). Brimley's mammal mounts gave the people of North Carolina a connection with their own vanished wildlife. Awe-struck visitors frequently leaned over the exhibit railing to touch the buffalo's shaggy fur, and in time the great buffalo mount was so damaged by the contact that it had to be removed.

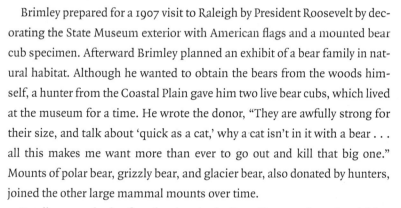

Brimley prepared for a 1907 visit to Raleigh by President Roosevelt by decorating the State Museum exterior with American flags and a mounted bear cub specimen. Afterward Brimley planned an exhibit of a bear family in natural habitat. Although he wanted to obtain the bears from the woods himself, a hunter from the Coastal Plain gave him two live bear cubs, which lived at the museum for a time. He wrote the donor, "They are awfully strong for their size, and talk about 'quick as a cat,' why a cat isn't in it with a bear . . . all this makes me want more than ever to go out and kill that big one." Mounts of polar bear, grizzly bear, and glacier bear, also donated by hunters, joined the other large mammal mounts over time.

Small mammals also found a place in the State Museum. In early exhibits, Brimley mounted fox squirrels and opossums in naturalistic poses on tree branches. H. H. Brimley's 1897 exhibit of an opossum family group is still on display, giving urban visitors a close-up look at wildlife. In the 1920s, folklorist and playwright Paul Green sought Brimley's advice on the state's opossum lore. Brimley replied, "Through the country districts in this section it has been quite accepted that the male possum copulates with the female through the nostrils, and that the birth of the young is also through the same openings." Brimley explained to the playwright this erroneous belief could have derived from opossum reproductive anatomy—female opossum's oviducts do not unite to form a uterus, and the male's penis is forked.

The State Museum no longer displays hunters' trophies; today habitat dioramas interpret the roles of mammals in natural communities. Since the 1960s, when David Adams first catalogued a disorganized quantity of mammal specimens dating to the 1800s, a systematics collection of mammals has supplied research material for the study of mammal species from manatees to moles. Insights into the lives of mammals big and small replace the tales of big game hunts that fascinated past generations.

Museum exterior decorated with bear mount for Teddy Roosevelt's 1907 visit (*above, right*); model in sporting attire posed with elk and buffalo mounts (*above, left*); gray wolf, *Canis lupus* (*opposite*). The sportsman's ethic ran strong through public life and museum exhibits in the early twentieth century. H. H. Brimley, an outdoorsman himself, evoked the fauna of the state's bygone wilderness by displaying mounts of large mammals that had been hunted to extinction in North Carolina—elk, buffalo, wolf, mountain lion. North Carolina State Museum Archives; wolf courtesy of Rosamond Purcell.

A Model of Nature

The State Museum's collections document the rich diversity of nature in North Carolina—currently estimated at 1,000 vertebrate species, 15,000 terrestrial and freshwater invertebrates, 6,000 plants, and 300 minerals. For many years, collections on display in the Geology Room, the Bird Hall, and the Mammal Gallery gave visitors to the museum a sense of the great diversity of the parts that make up the natural world.

Scientists use data from each collection to approximate, or model, life and its response to the environment in each of the state's physiographic regions. Such a model is always under revision, just as nature itself constantly changes. The scientists' model may take shape in publications, lectures, and documentaries. A literal model, a series of habitat dioramas based on information held in the collections, interprets nature for today's museum visitor. Habitat dioramas offer visitors a sense of how the parts interact as a whole—how biological populations have responded to geologic processes, changes in climate, geographic location, and altered habitats.

The Mountains

A series of continental collisions beginning more than a billion years ago created a chain of mountains at the eastern edge of North America. Much eroded by water and wind, the ancient southern Appalachians harbor an amazing diversity of flora and fauna. Tall ridges, with peaks over 6,500 feet

"Variety of form and habit is endless and when the student has spent a life in their examination he finds that he has hardly stepped upon the threshold of this wide world of wonders" (The Reverend Moses Ashley Curtis, 1834).

Unsorted eggshells. Courtesy of Rosamond Purcell.

in elevation, mark the eastern continental dividing line for river drainages. Five river basins in western North Carolina drain to the Gulf of Mexico. Twelve river basins flow into the Atlantic Ocean along the shores of the Carolinas. Eastern and western drainages have distinct faunas, especially of fishes and aquatic invertebrates. Almost a third of the state's crayfish species occur only in the five western river basins. Western fishes include the Kanawha darter and saffron shiner. The redlip shiner and pinewoods darter occur in eastern rivers.

The presence of some species is due to the region's geographical location. The New England cottontail rabbit is at the southern end of its range in the mountains, as are some shrews, bats, and other small mammals. Major flyways for migratory raptors and Neotropical songbirds traverse the Mountain region.

Geographical barriers—streams, deep valleys, mountaintops—created isolated populations, which gave rise to the current highly diverse salamander and millipede faunas. The many habitats in the region—peaks, coves, rocky outcrops, grass and heath balds, and spring-fed bogs—support a diversity of animals and plants. The bog turtle lives only in bogs of the mountains and upper Piedmont. The spotted skunk favors open woodlands, the mottled sculpin inhabits rocky stream bottoms, and the Yonahlossee salamander is found in mountain coves.

Cove forests, characterized by a tall tree canopy and moist, rich soils, are home to a huge diversity of species—more than 40 breeding birds, 100 trees, and 1,500 flowering plants. Coves contain the highest density of salamander species anywhere, and the southern Appalachians support 2,000 species of fungi.

Red spruce and Fraser fir evergreen trees, predominant at high elevations, are living reminders of a time when these hardy trees grew across much of the state. As the climate warmed around 10,000 years ago, warm-climate

(*above*)
Morel mushrooms. Fungi, wildflowers, and salamanders are among the many groups of organisms that have diversified in rich mountain cove habitat. Courtesy of Mike Dunn.

(*opposite*)
A great diversity of plants and animals is found in the mountain cove natural communities of the ancient southern Appalachians. Courtesy of Ken Taylor.

natural communities replaced spruce-fir forests. Only on North Carolina's highest peaks was the environment still cold enough for the plants and animals of the spruce-fir forest, including bird subspecies such as Cairns's black-throated blue warbler and the Appalachian black-capped chickadee. These "islands in the sky" are also home to the endemic Blue Ridge goldenrod, Weller's salamander, and the spruce-fir moss spider. The Carolina northern flying squirrel lives only in the spruce-fir forests of North Carolina and Tennessee.

Human activity, both inadvertent and intentional, has altered mountain populations of animals and plants. Big mammals were extirpated within the last two centuries—among them elk, buffalo, cougar, and wolf. Second-growth woods and clearings replaced old-growth forests, providing habitat for house wrens, meadowlarks, and brown-headed cowbirds, which in turn reduced habitat for native species. Native species compete with introduced species in streams, where the success of rainbow trout populations resulted in reduced populations of native brook trout.

Air pollution may be the cause of reduced growth in red spruce trees. A tiny introduced insect, the balsam woolly adelgid, has killed more than half of the region's Fraser fir trees since the 1960s, affecting untold numbers of plants and animals in spruce-fir communities.

The Piedmont

The hills of the Piedmont range in elevation from 300 to 1,500 feet. Deposits of gold and granites in the region are the result of a complex geologic past. Shale beds of Triassic rift basins preserve fossils hundreds of millions of years old, and clay fields made the region famous for brick production and pottery. With fertile soil and plentiful water resources in the twelve river basins that cut through the region, the Piedmont attracted the majority of settlers in the

Oak forests along Piedmont streams shelter migrating birds, small mammals and deer, and many reptiles and amphibians. Courtesy of John Connors.

state. The effects of 400 years of agriculture, stream damming, mining, and urbanization have radically altered the Piedmont's natural communities.

Along the streams of a Piedmont oak forest are plants and animals that have adapted surprisingly well to changes in the environment. Oak and hickory trees rise above red maples, dogwoods, and sourwoods, providing shelter for screech owls, red-tailed hawks, and songbirds. Green treefrogs, ground skinks, box turtles, and northern water snakes are fairly common. The fieryblack shiner, creek chub, speckled killifish, and striped jumrock populate Piedmont streams. While some mammals have been extirpated from the region—bear, cougar, forest bats—others thrive, including opossum, gray squirrel, and raccoon. The abundance of cleared land enabled the groundhog to move into the region from the mountains within the last 50 years, and also created habitat for nesting tree swallows. Farmland abandoned since the Depression in the 1930s is slowly returning to oak and hickory forest.

The marbled sala-mander, *Ambystoma opacum*, lives under rocks and logs in moist Piedmont habi-tats. Courtesy of Paris Trail.

Demands placed on land and water resources by agriculture and urban de-velopment disturb natural habitat in many areas, but some animals—in-sects, rodents, birds—may thrive where new food sources appear. Dams in Piedmont rivers affect aquatic species, including freshwater mussels and spawning fishes. Introduced game fishes like the flathead catfish compete with native fish like the robust redhorse. Non-native plants, including kudzu, wisteria, and Japanese honeysuckle, crowd out native plants in some areas. Populations of white-tailed deer and beaver may venture into suburban com-munities, vying with people for resources.

The Coastal Plain

The youngest region of the state, the Coastal Plain has been both below and above water as sea level fluctuated with the ice ages. Many fossils, including those of Cretaceous dinosaurs, Pliocene whales and walruses, and Pleis-tocene mastodons and ground sloths, are preserved in the Coastal Plain's sedimentary rock layers. Because of the region's geographic location, some southern species are found near the northern edges of their ranges here, in-cluding manatee, alligator, anhinga, Seminole bat, and Rafinesque's big-eared bat.

Fairly recent geologic events left unusual formations with distinctive habi-tats in North Carolina's Coastal Plain. Carolina bays, including Lake Wacca-maw, were most likely carved from sandstone by prevailing winds. The Wac-camaw drainage is home to crayfishes, fishes, and mollusks found nowhere else.

Longleaf pine forests covered much of the Coastal Plain before logging, agriculture, and fire suppression transformed the region. Only small patches

(*right, top*)
Fire-dependent long-leaf pine savannas support a distinctive fauna, including red-cockaded woodpeckers and fox squirrels. Courtesy of Mike Dunn.

(*right, bottom*)
A green treefrog, *Hyla cinerea*, finds a perch in a pitcher plant, one of 70 species of carnivorous plants found in wet savanna communities. Courtesy of David Lee.

(*opposite*)
Plants and animals have adapted to seasonal flooding in the hardwood bottomlands along the Roanoke River. Courtesy of Mike Dunn.

remain of the original 10-million-acre forest. Longleaf pines spread from the dry Sandhills region to the wet savannas of the lower Coastal Plain. These open, parklike forests depend on periodic fires, which leave bits of charcoal in the dark soil. Where no fires limit their growth, oak trees tower over and shade out young longleaf pines. Remaining longleaf pine habitats are havens for red-cockaded woodpeckers, gopher frogs, Bachman's sparrows, fox squirrels, and scarlet king snakes. Seasonal pools and an abundance of insects create ideal habitat for treefrogs. Wet pine savannas are home to the most diverse small-scale natural communities in the Western Hemisphere— 52 species have been counted in one square meter. Wildflowers, grasses, and carnivorous plants—Venus' flytraps, pitcher plants, and sundews—cover the ground in the rare wet pine savanna habitat.

Coastal Plain bottomland hardwood forests are found in river floodplains alongside agricultural and timber tracts. Oaks and other hardwoods shade the dry ridges, and water-tolerant bald cypress and gum trees spread in the

wet, low-lying areas. Waterlogged soil lacks enough oxygen to support decomposers, so rich organic matter colors the soil black. Plants and animals that live here must adapt to seasonal flooding, when heavy rains cause coastal rivers to overflow their banks. Remaining old-growth bottomland forests shelter huge trees with ample roosting cavities for forest bats.

Wildlife is abundant in the Coastal Plain. Bowfin and longnose gar fishes inhabit its sluggish creeks. Great blue herons and egrets enliven the wetlands, where Neotropical birds breed in spring and summer. Great flocks of overwintering waterfowl, decimated by hunting in the nineteenth century, have returned thanks to protective legislation, though habitat destruction has taken a toll on migratory populations. Diamondback terrapins rebounded with a decrease in hunting pressures, but are now threatened by habitat destruction around waterfronts. Restocking of wild turkey, quail, deer, and bear have boosted these game animal populations. Spawning fishes, including sturgeon, shad, and herring, have dwindled because dams on Coastal Plain rivers prevent their seasonal migration upstream. Introduced species, including crayfishes, fishes, and nutria (an aquatic mammal from South America), have entered ecosystems populated by native cottonmouths, snapping turtles, yellow-bellied turtles, and leopard frogs.

The rivers of the Coastal Plain mix fresh water with salt water in the estuaries. Nutrient-rich, brackish waters make these areas some of the most productive natural communities on Earth—communities that nurture the young of many sea animals, including 90 percent of all seafood species. In the salt marshes, wide prairies of *Juncus* and *Spartina* grasses channel meandering tidal creeks as they also trap sediments. Clams, snails, worms, fishes, and crabs are part of the marsh food web. In the open sounds, oyster reefs provide habitat for a wide diversity of animals. Water pollution in the region has decreased populations of fishes, arthropods (including shrimp and crab), and mollusks (including oysters).

(*left*)
Maritime forests of live oak and yaupon fringe one of the highest dunes on the East Coast, Jockey's Ridge in Dare County. Courtesy of Mike Dunn.

(*opposite*)
A great egret stalks its prey in a salt marsh, one of the most productive natural communities in North Carolina. Courtesy of Mike Dunn.

Separated from the mainland by stretches of sound waters up to 30 miles wide, barrier islands were created by climatic changes at the end of the last ice age. Once mainland sand ridges, they became islands when melting glaciers caused sea level to rise. Several natural communities exist on the barrier islands. Maritime forests feature some trees common to the mainland, like live oaks, loblolly pines, and cabbage palms, in addition to yaupons, wax myrtles, and cotton bushes. Gray squirrels, barn owls, glass lizards, and ospreys have adapted to life here. At Nags Head Woods, chicken turtles, oak toads, and pine woods snakes may be descendants of original mainland populations that were cut off when the Outer Banks islands formed.

Dune and grassland communities support few plants besides sea oats and dune spurge. Small animals like ant lions, tiger beetles, ghost crabs, and snakes may skitter across the sand. Baby loggerhead turtles hatch from their sandy nests on the beach in early fall. A few inches below the wet sand in the intertidal zone, mole crabs and coquina clams filter seawater for food. They in turn provide a meal for shorebirds. Rookeries support breeding birds by the thousands in spring, including ibises, herons, terns, pelicans, and gulls.

The Ocean

Life in the ocean is as much a product of climate, location, geologic process, and altered habitat as is life on land. The continental shelf, with its long, gradual slope, was above water during the last ice age. The rocky banks of old river channels, called hardbottoms, are now 60 to 110 feet below the ocean's surface. More than 300 kinds of fish live in these hardbottom communities, including tropical angelfish and butterflyfish, attracted by rich for-

(*above*)
Bird conservationist T. Gilbert Pearson studied pelicans roosting on the North Carolina coast in the 1910s. Data accumulated over time help scientists detect changes in populations of species like the pelican, once endangered but now rebounding in North Carolina. North Carolina State Museum Archives.

(*opposite*)
Pelicans are among the many bird species that nest in large colonies on the North Carolina coast. Courtesy of David Lee.

Short-billed marsh wren, *Cistothorus platensis*. Courtesy of David Lee.

aging grounds. Sea whips, worms, and algae attach to limestone ledges. Fish, crabs, and urchins fill every nook and cranny.

About 45 miles offshore on the edge of the continental slope, major oceanic currents meet: the Labrador current from the north and the subtropical waters of the Gulf Stream from the south. Fin whales, humpback whales, sperm whales, Atlantic leatherback turtles, loggerhead turtles, and birds from northern and southern hemispheres forage in these nutrient-rich waters. Degradation of breeding habitat in the Caribbean has taken its toll on bird species such as the black-capped petrel and Bermuda petrel, which spend part of their year here off the Carolina coast. With fewer suitable beaches for nesting, sea turtle populations have declined in the region. Hunting has greatly diminished whale populations, eliminating the Atlantic gray whale altogether and endangering the right whale and other great whales that swim in the state's waters.

All of life, from high mountains to deep seas, has been shaped by North Carolina's geographic location, long and eventful geologic history, varied climate, and human population. Early naturalists wrote of a land favored with unparalleled natural beauty and a great diversity of species. Today we know that diversity to be greater than they could have imagined.

A scientific understanding of nature—a model of the natural world—allows us to read danger signals when populations dip, or to judge when environmental factors tip out of balance. Based on a knowledge of nature, past and present, this model may help predict changes in the environment. With this knowledge comes a responsibility, felt by those who study and enjoy the natural world in all its complexity. The power to affect the future of that world is in our hands.

SOURCES

One. Collecting Nature

Much of the information on the history of the State Museum came from extensive research conducted by Eloise Potter for a manuscript entitled *The North Carolina State Museum of Natural Sciences, 1879–1990*. Copies of the manuscript are available in the museum's library. The text explores the role played by the museum in the educational efforts of the N.C. Department of Agriculture. Raymond Beck, state capitol historian, and William Palmer and John Cooper, NCSM curators, were knowledgeable sources on the museum's roots. Ms. Potter, Robert Wolk, and Stephen Busack, also of NCSM, provided insight into the history of natural history studies. Wayne Martin and Beverly Patterson of the North Carolina Arts Council provided a transcript of their 1994 interview with the Reverend Robert Bushyhead.

Sources that pertain to multiple chapters are cited only in the first chapter.

NCSM = North Carolina State Museum of Natural Sciences.

North Carolina State Board of Agriculture. *North Carolina and Its Resources*. Raleigh: N.C. State Board of Agriculture, 1896.

Allmon, Warren D. "The Value of Natural History Collections." *Curator* 37, no. 2 (1994):83–89.

Barber, Lynn. *The Heyday of Natural History 1820–1870*. Garden City, N.Y.: Doubleday and Co., 1980.

Bartram, William. *The Travels of William Bartram*. 1791. Francis Harper's Naturalist Edition. Athens: University of Georgia Press, 1998.

Bolen, Eric. "The Bartrams in North Carolina." *Wildlife in North Carolina* (May 1996):16–21.

———. "John Lawson's Legendary Journey." *Wildlife in North Carolina* (December 1998):22–27.

Brimley, H. H. *Hand Book of the North Carolina State Museum*. Raleigh: North Carolina State Board of Agriculture, 1897.

Barking treefrog skeleton, *Hyla gratiosa*. Courtesy of Rosamond Purcell.

Cooper, John. "NC State Museum of Natural History." *Association of Systematics Collections Newsletter* 10, no. 3 (June 1982):25–28.

———. "The Brothers Brimley." *Brimleyana* 1 (1979):1–14.

Farber, Paul Lawrence. *Finding Order in Nature.* Baltimore: Johns Hopkins University Press, 2000.

Feduccia, Alan, ed. *Catesby's Birds of Colonial America.* Chapel Hill: University of North Carolina Press, 1985.

Hale, P. M. *The Woods and Timbers of North Carolina.* Raleigh: Edwards and Broughton, 1890.

Hariot, Thomas. *A Briefe and True Report of the New Found Land of Virginia.* 1588. Online archive edition by Melissa S. Kennedy. University of Virginia, 1996. <http://www.people.virginia.edu/ffimsk5d/hariot/main.html>.

Hubbell, Sue. "The New Taxonomy: Nothing Like You Learned in School." *Smithsonian Magazine* (May 1996):140–51.

Lawson, John. *A New Voyage to Carolina.* Edited by Hugh Lefler. Chapel Hill: University of North Carolina Press, 1967.

Mooney, James. *James Mooney's History, Myths, and Sacred Formulas of the Cherokees.* Asheville, N.C.: Bright Mountain Books, 1992.

Odum, Eugene. *A North Carolina Naturalist, H. H. Brimley.* Chapel Hill: University of North Carolina Press, 1949.

Saunders, John R. *The World of Natural History.* New York: Sheridan House, 1952.

Schwarzkopf, S. Kent. *A History of Mt. Mitchell and the Black Mountains.* Raleigh: North Carolina Department of Cultural Resources, 1985.

Simpson, Marcus, and Sallie Simpson. "Moses Ashley Curtis (1808–1872)." *North Carolina Historical Review* 60, no. 2:137–70.

Sterling, Elizabeth. "History of the Bug House Laboratory, Washington, NC." Manuscript.

Wilson, E. O. *The Diversity of Life.* New York: W.W. Norton, 1992.

Wonders, Karen. *Habitat Dioramas: Illusions of Wilderness in Museums of Natural History.* Acta Universitatis Upsaliensis. Figura Nova Series 25. Upsalla: University of Upsalla Press, 1993.

Two. Rocks and Minerals

Chris Tacker, curator in geology at NCSM, offered a wealth of ideas, information, and direction. Phil Cox, NCSM, and Jeff Reid and Tyler Clark of the North Carolina Geological Survey supplied valuable information. Gail Gillespie advised on the history of mineral springs in North Carolina.

Beyer, Fred. *North Carolina: The Years Before Man.* Durham, North Carolina: Carolina Academic Press, 1991.

Cabe, Suellen. "Cretaceous and Cenozoic Stratigraphy of the Upper and Middle Coastal Plain, Harnett County Area, NC." Ph.D. diss., University of North Carolina, 1984.

Carpenter, P. Albert. *A Geologic Guide to North Carolina's State Parks.* Raleigh: North Carolina Geologic Survey, 1989.

Emmons, Ebenezer. *Geological Report of the Midland Counties of North Carolina.* New York: Putnam, 1856.

———. *Report of the N.C. Geological Survey (With Descriptions of the Fossils of the Marl Beds).* Raleigh: H.D. Turner, 1858.

Jeffrey, Thomas E. *Thomas Lanier Clingman, Fire Eater from the Carolina Mountains.* Athens: University of Georgia Press, 1998.

Kerr, W. C. *Report of the Geological Survey of North Carolina.* Vol. 1. Raleigh: Josiah Turner, 1875.

Mitchell, Elisha. *Elements of Geology.* Chapel Hill: State of North Carolina, 1842.

Stuckey, J. L. *North Carolina: Its Geology and Mineral Resources.* Raleigh: North Carolina Department of Conservation and Development, 1965.

Zug, Charles G. *Turners and Burners.* Chapel Hill: University of North Carolina Press, 1998.

Three. Fossils

Vince Schneider, curator in paleontology, and Patricia Weaver, collections manager, NCSM, assisted in the development of this book by advising on direction, organization, and clarity. Dale Russell, senior curator in paleontology, was especially helpful in describing the Cretaceous world of the dinosaurs.

Baird, Donald, and John R. Horner. "Cretaceous Dinosaurs of North Carolina." *Brimleyana,* no. 2 (November 1979):1–28.

Carter, J. G., P. E. Gallagher, R. Enos Valone, and T. J. Rossbach. *Fossil Collecting in North Carolina.* Bulletin 89. Raleigh: N.C. Department of Natural Resources and Community Development, 1988.

Emmons, Ebenezer. *Report of Professor Emmons on His Geological Survey of North Carolina.* Raleigh: Seaton Gales, 1852.

Fisher, Paul E., D. A. Russell, M. K. Stoskopf, R. E. Barrick, M. Hammer, A. A. Kuzmitz. "Cardiovascular Evidence for an Intermediate or Higher Metabolic Rate in an Ornithischian Dinosaur." *Science* 288 (2000):503–5.

Johnson, Kirk R., and Richard K. Stucky. *Prehistoric Journey.* Boulder, Colo.: Roberts Rinehart, 1995.

Lee, David. "Dinosaurs Down the Cape Fear." *Wildlife in North Carolina* (April 1994):20–25.

Lyell, Charles. *Travels in North America.* Vol. 1. New York: Wiley and Putnam, 1845.

Monastersky, Richard. "The Rise of Life on Earth" *National Geographic* (March 1998):54–81.

Payne, Peggy. "Casting Into an Ancient Sea." *Wildlife in North Carolina* (April 1980):8–13.

Potter, Eloise. "Nature Notes: Fossils in North Carolina." Raleigh: NCSM, 1993.

Pratt, Joseph Hyde. *Report of the N.C. Geological and Economic Survey.* Vol 9. Raleigh: State Printer and Binder, 1912.

Russell, Dale. *An Odyssey in Time: Dinosaurs of North America.* Toronto: University of Toronto Press, 1989.

———. Dinosaurs Down South. *N.C. Naturalist* (Autumn 1996):1–2.

Sandhu, Pavi. "Plant Fossils Yield Clues to Climate." *News and Observer.* 12 August 1997.

Schneider, Vince, and Robert G. Wolk. "Nature Notes: Giant Ground Sloth." Raleigh: NCSM, 1997.

Simpson, George. *Proceedings of the American Philosophical Society.* 86 (1942):130–188.

Vanderbilt, Tom. "Rock Solid Find." *Concord Tribune.* 26 November 1997.

Four. Invertebrates

Rowland Shelley, curator of terrestrial invertebrates, John Cooper, curator of crustaceans, and Arthur Bogan, curator of aquatic invertebrates, generously gave of their time and knowledge in shaping the wide-ranging invertebrates story. Also helpful were Jesse Perry and Bill Reynolds, NCSM, Rachel Yahn and Louise Benner of the North Carolina Museum of History, archeologists Tom Hargrove, Mark Mathis, and Ned Woodall, and entomologist David Stephan, North Carolina State University.

Shelley, Rowland. "History of the Invertebrate Collection and Research Program (through 1996)." Manuscript, NCSM.

MILLIPEDES

Lee, David. "Dr. Millipede." *Wildlife in North Carolina* (May 1996):12–15.

Shelley, Rowland. "Centipedes and Millipedes." Kansas School Naturalist 45, no. 2 (October 1998):1–13.

———. "Hundred-leggers, Thousand-leggers." *Wildlife in North Carolina* (September 1976):23–25.

Whitehead, Donald, and Rowland Shelley. "Mimicry among Aposematic Appalachian Xystodesmid Millipeds." *Proceedings of the Entomological Society of Washington* 94, no. 2 (1992):177–88.

INSECTS

Brimley, C. S. *Insects of North Carolina.* Raleigh: North Carolina Department of Agriculture, 1938.

Ellis, Harry. "The World is Full of Moths." *Wildlife in North Carolina* (November 1998):14–17.

Peigler, Richard. "Moth Cocoon Artifacts." *Cultural Entomology Digest* (November 1997). <http:///www.insects.org/ced4/peigler.html>

CRAYFISH

Cooper, John E., and Alvin Braswell. "Observations on North Carolina Crayfishes." *Brimleyana* (22 June 1995):87–132.

Cooper, John E. and M. R. Cooper. "A New Species of Crayfish of the Genus *Orconectes,* Subgenus *Procericambarus* (Decapoda: Cambaridae) Endemic to the Neuse and Tar-Pamlico River Basins, NC." *Brimleyana* 23 (December 1995):65–87.

Cooper, John E., A. L. Braswell, and C. McGrath. "Noteworthy Distributional Records for Crayfishes in North Carolina." *Journal of the Elisha Mitchell Scientific Society* 114, no. 1 (1998):1–10.

Cooper, John E. "A New Species of Crayfish of the Genus *Procambarus,* Subgenus *Ortmannicus* (Decapoda: Cambaridae), from the Waccamaw River Basin, North and South Carolina." *Proceedings of the Biological Society of Washington* 111, no.1 (1998):81–91.

Ellis, Harry. "Discovering Crayfish." *Wildlife in North Carolina* (June 1998):24–27.

Stager, J. C., and L. B. Cahoon. "The Age and Trophic History of Lake Waccamaw, NC." *Journal of the Elisha Mitchell Scientific Society* 103, no. 1 (1987):1–13.

LEECHES

Horan, Jack. "Leeches Latch on to Medical Mainstream." *Charlotte Observer.* 8 March 1996.

Shelley, Rowland. "North Carolina's Terrestrial Leech." *Wildlife in North Carolina* (September 1977):24.

Shelley, Rowland, Alvin Braswell, and David Stephan. "Notes on the Natural History of the Terrestrial Leech, *Haemopis septagon.*" *Brimleyana* 1 (March 1979):129–33.

Bogan, Arthur. "Freshwater Bivalve Extinctions (Mollusca: Unionoida): A Search for Causes." *American Zoologist* 33 (1993):599–609.

———. "The Silent Extinction." *American Paleontologist* 5, no.1 (February 1997):2–4.

Christmas, John, K. Morivans, and A. Bogan. "Freshwater Unionid Mussel in Maryland." *Maryland Fish and Wildlife News* 5, no. 4 (Summer 1997):2.

Shelley, Rowland. "In Defense of Naiades." *Wildlife in North Carolina* (March 1972):4–8.

Earley, Lawrence. "Declaration of Interdependence." *Wildlife in North Carolina* (June 1998):2–3.

Five. Fishes

Wayne Starnes, curator of fishes, NCSM, unselfishly lent to this project his superb knowledge of southeastern fishes, his collection of rare manuscripts, and excellent editorial advice. Gabriela Hogue, collection manager, NCSM, and technician Lynn Womack provided willing support. The fishes collection has been served by curators William Palmer, Alvin Braswell, and Frank Snelson, and by collectors for periodic fish surveys conducted by state and federal agencies, including Vince Schneider. In addition to the original collection begun in the 1800s, the museum's fishes collection comprises collections donated by the Institute of Marine Sciences–University of North Carolina, Duke University, North Carolina State University, Mars Hill College, Stockton State College (New Jersey), and the personal collections of Wayne Starnes, Fred C. Rohde, and Rudolph G. Arndt.

Bryant, Richard, J. W. Evans, R. E. Jenkins, and B. J. Freeman. "The Mystery Fish (Robust Redhorse)." *Southern Wildlife* 1, no. 2 (1996):26–35.

Cope, Edward D. "A Partial Synopsis of the Fishes of the Fresh Waters of North Carolina." *American Philosophical Society* 11 (1870):448–95.

Early, Lawrence. *Nature's Ways, Volumes 1–6*. Raleigh: N.C. Wildlife Resources Commission, 1997.

Etnier, David A., and Wayne C. Starnes. *The Fishes of Tennessee*. Knoxville, University of Tennessee Press, 1993.

Evermann, Barton W., and Ulysses O. Cox. "The Fishes of the Neuse River Basin." *Bulletin of the U.S. Fish Commission* 10 (1890, published 1892):303–7.

Hubbs, Carl L., and Edward C. Raney. *Endemic Fish Fauna of Lake Waccamaw, NC*. Ann Arbor: University of Michigan Press, 1946.

Jordan, David Starr. *The Days of a Man*. Vol. 1. Yonkers-on-Hudson, N.Y.: World Book Co., 1922.

———. *Fishes*. London: D. Appleton and Co., 1925.

Kemp, Karen. "Time in a Bottle." *North Carolina Naturalist* (Spring 1997):2–9.

Lee, David, C. P. Gilbert, C. H. Hocutt, R. E. Jenkins, D. E. McAllister, and J. R. Stauffer Jr. *Atlas of North American Freshwater Fishes*. Raleigh: North Carolina State Museum of Natural History, 1980.

Lee, David. "Deep Sea Slime Machine." *Wildlife in North Carolina* (August 1998):2–3.

———. "Ecosystem of Castaways." *Wildlife in North Carolina* (August 1999):12–17.

Menhinick, Edward F., and Alvin L. Braswell. *Endangered, Threatened, and Rare Fauna of North Carolina. Part IV: A Reevaluation of the Freshwater Fishes*. Occasional Papers, no. 11. Raleigh: North Carolina State Museum of Natural Sciences, 1997.

Orr, Douglas. *North Carolina Atlas*. Chapel Hill: University of North Carolina Press, 1999.

Rohde, Fred, R. G. Arndt, D. G. Lindquist, and J. F. Parnell. *Freshwater Fishes of the*

Carolinas, Virginia, Maryland, and Delaware. Chapel Hill: University of North Carolina Press, 1994.

Smith, Hugh M. *The Fishes of North Carolina.* Vol. 2. Raleigh: North Carolina Geologic and Economic Survey, 1907.

Six. Reptiles and Amphibians

Curators Alvin Braswell and William Palmer, and collections manager Jeff Beane have contributed much to the herpetology collection of the State Museum and to the direction and accuracy of this chapter. Dennis Herman, coordinator of living collections, contributed research on the bog turtle.

Adler, Kraig. *Contributions to the History of Herpetology.* St. Louis, Mo.: Society for the Study of Amphibians and Reptiles, 1989.

————. *Herpetology in North America before 1900.* Circular no. 8. Society for the Study of Amphibians and Reptiles, Lawrence, Ks.: University of Kansas, 1979.

Braswell, Alvin. "Survey of the Amphibians and Reptiles of Nags Head Woods Ecological Preserve." *Association of Southeastern Biologists Bulletin* 35, no. 4 (October 1988):199–217.

————. "Status Report on *Rana capito capito* LeConte, the Carolina Gopher Frog in N.C." Report to N.C. Wildlife Resources Commission, January 1993.

Cooper, John, and Ray Ashton. "The *Necturus lewisi* Study: Introduction, Selected Literature Review, and Comments on the Hydrologic Units and Their Faunas." *Brimleyana* 10 (February 1985):1–12.

Dopyera, Caroline. "The Shrinking Pool." *News and Observer.* 23 January 1994.

Kemp, Karen. "Turtle on a Tightrope." *North Carolina Naturalist* 5, no. 2 (1997):2–9.

Martof, Bernard S., W. Palmer, J. Bailey, and J. Harrison. *Amphibians and Reptiles of the Carolinas and Virginia.* Chapel Hill: University of North Carolina Press, 1980.

Palmer, William, and Alvin Braswell. *Reptiles of North Carolina.* Chapel Hill: University of North Carolina Press, 1995.

Palmer, William. *Poisonous Snakes of North Carolina.* Raleigh: North Carolina State Museum of Natural History, 1974.

Sterling, Elizabeth. "History of the Bug House Laboratory, Washington, NC." Manuscript.

Seven. Birds

David Lee, curator of birds, John Gerwin, collections manager, and technician Becky Browning, NCSM, contributed unstintingly of their time and expertise in the field of southeastern ornithology. David contributed greatly from his own extensive research and suggested appropriate archival materials.

Lee, David S. "Joyce Kilmer's Birds." *Wildlife in North Carolina* (October 1997):4–9.

————. "Extinction, Extirpation, and Range Reduction of Breeding Birds in North Carolina: What Can Learned?" *Chat* 63, no. 3 (1999):103–22.

————. "Range Expansion of the Tree Swallow in the Southeastern U.S." *Brimleyana* 18 (1993):103–113.

————. "Pelagic Seabirds and the Proposed Exploration for Fossil Fuels Off North Carolina: A Test for Conservation Efforts of a Vulnerable International Resource." *Journal of the Elisha Mitchell Scientific Society* 115, no. 4 (1999):294–315.

Lee, David S., and Becky Browning. "Conservation Concerns Related to Avian Endemism in Southern Appalachians." Manuscript.

Potter, Eloise, J. F. Parnell, and R. Teulings. *Birds of the Carolinas*. Chapel Hill: University of North Carolina Press, 1980.

Pearson, T. Gilbert, C. S. Brimley, and H. H. Brimley. *Birds of North Carolina*. Raleigh: North Carolina Department of Agriculture, 1959.

Simpson, Marcus. *Birds of the Blue Ridge*. Chapel Hill: University of North Carolina Press, 1992.

Eight. Mammals

Mary Kay Clark, curator of mammals, NCSM, offered valuable advice not only on the subject of mammalogy, but on the meaning and purpose of collections. Collections manager Lisa Gatens helped ensure the accuracy of the text, and John Kelly and Kate Marsh rendered assistance in the mammals collection. Barbara Falzone and Bob Alderink, also NCSM, offered useful information.

Adams, David A. "Changes in Forest Vegetation, Bird, and Small Mammal Populations at Mount Mitchell, NC—1959/62 and 1985." *Journal of Elisha Mitchell Scientific Society* 107, no. 1 (1991):3–12.

Brown, Lawrence. *A Guide to the Mammals of the Southeastern United States*. Knoxville: University of Tennessee Press, 1997.

Carwardine, Mark. *Whales, Dolphins, and Porpoises*. London: Dorling Kindersley, 1995.

Clark, Mary Kay, ed. *Endangered, Threatened and Rare Fauna of North Carolina. Part 1: A Reevaluation of the Mammals*. Occasional Papers, no. 3. Raleigh: North Carolina State Museum of Natural Sciences, 1987.

———. "Skull Sessions." *Wildlife in North Carolina* (November 1997):12–17.

Faris, Jeannie. "Old Forests May Be Last Refuge for Rare Bat." *Coastwatch* (March/April 1994):16–21.

Graham, Frank, Jr. "In Search of Beidler's Bats." *Audubon* (May-June 1995):106–9.

Kemp, Karen. "Stalking Bats." *North Carolina Naturalist* (Fall/Winter 1999):2–7.

Lee, David, John B. Funderburg Jr., and Mary K. Clark. *A Distributional Survey of North Carolina Mammals*. Occasional Papers, no. 10. Raleigh: North Carolina State Museum of Natural History, 1982.

Simpson, Marcus B., and Sallie W. Simpson. *Whaling on the North Carolina Coast*. Raleigh: North Carolina Department of Cultural Resources, 1990.

Webster, William D., J. Parnell, and W. Briggs. *Mammals of the Carolinas, Virginia, and Maryland*. Chapel Hill: University of North Carolina Press, 1985.

Nine. A Model of Nature

Mike Dunn and Alvin Braswell, NCSM, were most helpful in organizing this material, as were all exhibit text contributors to the museum's habitat exhibits—Mountains to the Sea and Coastal North Carolina.

Schafale, Michael P., and Alan S. Weakley. *Classification of the Natural Communities of North Carolina (Third Approximation)*. Raleigh: Natural Heritage Program, North Carolina Department of Environment, Health, and Natural Resources, 1990.

Early, Lawrence, ed. *North Carolina Wild Places*. Raleigh: North Carolina Wildlife Resources Commission, 1993.

INDEX

Page numbers in italics refer to illustrations.

Mineral cores from nineteenth-century North Carolina mine. Courtesy of Rosamond Purcell.